Methods in Molecular Biology

Series Editor
John M. Walker
School of Life and Medical Sciences
University of Hertfordshire
Hatfield, Hertfordshire, AL10 9AB, UK

For further volumes:
http://www.springer.com/series/7651

Chimera Research

Methods and Protocols

Edited by

Insoo Hyun

Department of Bioethics, Case Western Reserve University, School of Medicine, Cleveland, OH, USA

Alejandro De Los Angeles

Department of Psychiatry, Yale University School of Medicine, New Haven, CT, USA

Editors
Insoo Hyun
Department of Bioethics
Case Western Reserve University
School of Medicine
Cleveland, OH, USA

Alejandro De Los Angeles
Department of Psychiatry
Yale University School of Medicine
New Haven, CT, USA

ISSN 1064-3745 ISSN 1940-6029 (electronic)
Methods in Molecular Biology
ISBN 978-1-4939-9526-4 ISBN 978-1-4939-9524-0 (eBook)
https://doi.org/10.1007/978-1-4939-9524-0

Cover illustration: Chimera Period: Six Dynasties (220 - 589) Date: 5th century Culture: China Medium: Stone Classification: Sculpture Credit Line: Fletcher Fund, 1973 Accession Number: 63.224.1 Dimensions: H. 21 1/2 in (54.6 cm); W. 18 in (45.7 cm); L. 18 1/2 in (47 cm).

This Humana imprint is published by the registered company Springer Science+Business Media, LLC, part of Springer Nature.
The registered company address is: 233 Spring Street, New York, NY 10013, U.S.A.

Preface

In the ancient imagination, the chimera was a monster comprised of parts from a lion, a goat, and a serpent. In modern biomedical research, chimeras are entities made up of cells from two or more zygotes of the same or different species. Experimental chimeras comprised of cells from two individuals, particularly in the mouse, are widely used in everyday biomedical research for generating transgenic mice. More recently, however, advances in the generation of chimera-competent pluripotent stem cells and interspecies chimera research are blazing new paths for applications of chimeras for basic biology and regenerative medicine. Generating human-animal chimeras using patient stem cells might create an in vivo setting to study human disease and to generate transplantable human organs inside large animals.

At present, there exist many questions surrounding chimeras. Interspecies chimeras manifest lower levels of donor chimerism when compared with intraspecies chimeras, suggesting the existence of a barrier to interspecies chimera formation. With specific regard to human-animal chimeras, currently available data show that the degree of human donor cell contribution to animal host embryos is very low. In the instances where donor cell engraftment may have occurred, the functionality of such cells is unclear. Therefore, one important question is whether more extensive human-animal chimerism can be achieved. Moreover, if human organs can be generated in human-pig chimeras, will such organs be transplantable given that host blood vessels and nerve cells may still be present?

For the promise of chimera research to be fully realized, the current limitations of interspecies chimeras need to be more thoroughly explored. These include understanding malformations, developmental arrest, and organ-to-organ variation in levels of donor chimerism. Further development of strategies will be needed to enhance the degree of human-animal chimerism. We need to understand the constituents of the species barrier that inhibit efficient colonization of animal embryos with human cells. It is clear that matching developmental speed between human donor cells and host animal cells will be needed to achieve coordinated morphogenesis and organogenesis. It may be necessary to deploy strategies to enhance the capabilities of human cells to compete equally with host cells. Finally, lowering human-animal interspecies barriers is likely to require "humanization" of large animal hosts by genetic engineering approaches.

This volume addresses provocative new questions surrounding stem cell-based chimera research, divided into three parts. In Part I, the book provides a summary of different human donor cell types. Alejandro De Los Angeles revisits the ever-evolving spectrum of pluripotency and provides a perspective on evaluating new types of pluripotent stem cells. De Los Angeles and Jun Wu also describe how to derive naïve and primed pluripotent stem cells from mouse preimplantation and postimplantation embryos. Rio Sugimura describes how to generate engraftable hematopoietic stem progenitor cells from human pluripotent stem cells. Wai Leong Tam and colleagues provide an overview of cancer cell biology. And Insoo Hyun explains the ethical and regulatory intricacies of informed consent for the procurement of somatic cells used to derive pluripotent stem cell lines utilized subsequently in chimera studies.

In Part II, the book provides various methods for generating chimeras, including those between human donor cells and nonhuman hosts. Ali Brivanlou and colleagues share their experimental protocols for chick models and human-chick organizer grafts. Byoung Ryu describes methods for transplanting human CD34+ cells into humanized mice. Juan Carlos Izpisua Belmonte and colleagues describe their methodology for generating human-pig interspecies chimeras. De Los Angeles and Wu present methods for generating embryonic chimeras between human and nonhuman primate pluripotent stem cells and mouse host embryos. De Los Angeles also shares experimental techniques for transplanting mouse neural stem cells into a mouse disease model for stroke. Hyun concludes Part II with a discussion of ethical standards for chimera research oversight.

In Part III, the book concludes by offering perspectives on ethical controversies and new scientific directions. Sebastian Porsdam Mann and others meditate on the ethics of crossing the xenobarrier. Daniel Counihan highlights the importance of animal welfare as an ethical consideration alongside concerns about the moral status of chimeras. Ralph Brinster and colleagues describe their methods for experimentation with spermatogonial stem cells. And De Los Angeles, Hyun, and colleagues recommend a cautious exploration of human-monkey chimera studies to further our understanding of complex human brain disorders.

Collectively, the chapters in this volume serve as a valuable resource for scientists interested in using chimeras as a research tool while appreciating their complex ethical dimensions. The path ahead has many challenges—scientific, medical, and ethical. The scientific community is obligated to approach these challenges and proceed within ethical guidelines.

Cleveland, OH, USA *Insoo Hyun*
New Haven, CT, USA *Alejandro De Los Angeles*

Contents

Contributors

RALPH L. BRINSTER • *Department of Biomedical Sciences, School of Veterinary Medicine, University of Pennsylvania, Philadelphia, PA, USA*
ALI H. BRIVANLOU • *Laboratory of Molecular Vertebrate Embryology, The Rockefeller University, New York, NY, USA*
DANIEL COUNIHAN • *Department of Bioethics, Case Western Reserve University, School of Medicine, Cleveland, OH, USA*
ALEJANDRO DE LOS ANGELES • *Department of Psychiatry, Yale University School of Medicine, New Haven, CT, USA*
JOHN D. ELSWORTH • *Department of Psychiatry, Yale University School of Medicine, New Haven, CT, USA*
RAJSHEKHAR R. GIRADDI • *Salk Institute for Biological Sciences, La Jolla, CA, USA*
GÖRAN HERMERÉN • *Department of Medicine, Lund University, Lund, Sweden*
INSOO HYUN • *Department of Bioethics, Case Western Reserve University, School of Medicine, Cleveland, OH, USA*
JUAN CARLOS IZPISUA BELMONTE • *Salk Institute for Biological Studies, La Jolla, CA, USA*
TATIANE Y. KANNO • *Laboratory of Molecular Vertebrate Embryology, The Rockefeller University, New York, NY, USA*
YOON-SANG KIM • *Department of Hematology, St. Jude Children's Research Hospital, Memphis, TN, USA*
STEPHEN R. LATHAM • *Yale Interdisciplinary Center for Bioethics, Yale University, New Haven, CT, USA*
MAY YIN LEE • *Genome Institute of Singapore, Singapore, Singapore*
SEBASTIAN PORSDAM MANN • *Department of Media, Cognition and Communication, University of Copenhagen, Copenhagen, Denmark; Uehiro Center for Practical Ethics, Oxford University, Oxford, UK*
IAIN MARTYN • *Laboratory of Molecular Vertebrate Embryology, The Rockefeller University, New York, NY, USA; Center for Studies in Physics and Biology, The Rockefeller University, New York, NY, USA*
DAIJI OKAMURA • *Department of Advanced Bioscience, Graduate School of Agriculture, Kindai University, Nara, Japan*
D. EUGENE REDMOND JR • *Axion Research Foundation, Hamden, CT, USA*
BYOUNG RYU • *Department of Hematology, St. Jude Children's Research Hospital, Memphis, TN, USA*
MASAHIRO SAKURAI • *Department of Molecular Biology, University of Texas Southwestern Medical Center, Dallas, TX, USA*
NILAM SINHA • *Department of Biomedical Sciences, School of Veterinary Medicine, University of Pennsylvania, Philadelphia, PA, USA*
RYOHICHI SUGIMURA • *Center for iPS Cell Research and Application, Kyoto, Japan*
ROSA SUN • *Nuffield Department of Clinical Neurosciences, University of Oxford, Oxford, UK; Department of Neurosurgery, Birmingham Hospital, Birmingham, UK*
WAI LEONG TAM • *Genome Institute of Singapore, Singapore, Singapore; Cancer Science Institute of Singapore, Singapore, Singapore; Yong Loo Lin School of Medicine, National*

University of Singapore, Singapore, Singapore; School of Biological Sciences, Nanyang Technological University, Singapore, Singapore

EOIN C. WHELAN • *Department of Biomedical Sciences, School of Veterinary Medicine, University of Pennsylvania, Philadelphia, PA, USA*

MATTHEW WIELGOSZ • *Department of Hematology, St. Jude Children's Research Hospital, Memphis, TN, USA*

JUN WU • *Department of Molecular Biology, University of Texas Southwestern Medical Center, Dallas, TX, USA; Hamon Center for Regenerative Science and Medicine, University of Texas Southwestern Medical Center, Dallas, TX, USA*

CUIQING ZHONG • *Salk Institute for Biological Studies, La Jolla, CA, USA*

Part I

Human Donor Cell Types: Potency and Position

Chapter 1

Frontiers of Pluripotency

Alejandro De Los Angeles

Abstract

Humans develop from a unique group of pluripotent cells in early embryos that can produce all cells of the human body. While pluripotency is only transiently manifest in the embryo, scientists have identified conditions that sustain pluripotency indefinitely in the laboratory. Pluripotency is not a monolithic entity, however, but rather comprises a spectrum of different cellular states. Questions regarding the scientific value of examining the continuum of pluripotent stem (PS) cell states have gained increased significance in light of attempts to generate interspecies chimeras between humans and animals. In this chapter, I review our ever-evolving understanding of the continuum of pluripotency. Historically, the discovery of two different PS cell states in mice fostered a general conception of pluripotency comprised of two distinct attractor states: naïve and primed. Naïve pluripotency has been defined by competence to form germline chimeras and governance by unique KLF-based transcription factor (TF) circuitry, whereas primed state is distinguished by an inability to generate chimeras and alternative TF regulation. However, the discovery of many alternative PS cell states challenges the concept of pluripotency as a binary property. Moreover, it remains unclear whether the current molecular criteria used to classify human naïve-like pluripotency also identify human chimera-competent PS cells. Therefore, I examine the pluripotency continuum more closely in light of recent advances in PS cell research and human interspecies chimera research.

Key words Pluripotency, Pluripotent stem cells, Naïve pluripotent, Stem cells, Naïve pluripotency, Primed pluripotent stem cells, Primed pluripotency, Formative pluripotency, Interspecies chimeras

1 Introduction

PS cells can be derived from a number of different sources [1]. The first diploid PS cells, mouse embryonic stem (ES) cells, were derived from the inner cell mass (ICM) of preimplantation embryos [2, 3]. Cell lines with pluripotency can also be generated from more developmentally advanced postimplantation epiblasts and such stem cells are known as epiblast stem (EpiS) cells [4, 5]. Most strikingly, an ES cell-like cell type, induced pluripotent stem (iPS) cells, can be derived from somatic cells by overexpressing the transcription factors (TFs) Oct4, Sox2, Klf4, and C-Myc in somatic cells [6]. Human and nonhuman primate (NHP) PS cells have also been derived from embryonic and somatic cells [7–9]. Patient-

Insoo Hyun and Alejandro De Los Angeles (eds.), *Chimera Research: Methods and Protocols*, Methods in Molecular Biology, vol. 2005, https://doi.org/10.1007/978-1-4939-9524-0_1,

specific iPS cells are anticipated to significantly contribute to disease research and medicine [10]. Although the defining properties of PS cells have been extensively reviewed by several groups [1, 11], our understanding of pluripotency is constantly evolving and in need of refinement. Consequently, it is imperative to revisit our ever-evolving understanding of PS cells because of their game-changing potential for regenerative medicine.

A few years ago, in the wake of the STAP scandal, laboratories around the world came together and provided guidelines for evaluating novel claims of pluripotency and defined the hallmarks of pluripotency [1]. In light of advances in PS cell research and interspecies chimera research, there exists a need to revisit the relevance of these markers. In this chapter, I provide a partial update to the Hallmarks of Pluripotency review [1].

2 Core Features of Pluripotent Stem Cells

2.1 Functional Characteristics

The cellular potential of single cells or aggregates of cells is measured through different functional assays (Fig. 1) [1, 11]. While assays for evaluating differentiation potential of PS cells have been reviewed previously [1, 11], each of these functional assays possesses nuances, which I will review in further detail below.

PS cells are defined by two properties: indefinite self-renewal in the laboratory and differentiation capacity to form the three embryonic germ layers—ectoderm, mesoderm, and endoderm. The differentiation potential of PS cells is measured by various functional assays, which include differentiation in the dish and production of tumors containing the three germ layers known as teratomas [11]. The production of chimeras—animals comprised of cells from two zygotes—is another major type of experiment for assessing stem cell potency. Experiments that demonstrate high cellular potency show germline transmission, tetraploid complementation, and chimera generation from single cells.

In vitro differentiation to derivatives of the three embryonic germ layers is arguably the most fundamental functional assay to assess for pluripotency. Differentiation of PS cells in vitro can be achieved by three different types of methods. The first type of differentiation protocol is to replace pluripotency maintenance culture conditions with mixtures of differentiation-inducing cytokines, morphogens, and/or chemicals on different extracellular matrices. The factors that are employed to trigger differentiation to a desired lineage have been identified in large part by studying mechanisms that regulate ontogeny or by screening libraries of small molecules for their function in induction of different fates. Another method for inducing differentiation in the dish is to generate embryoid bodies by transfer of two-dimensional (2D) PS cell cultures into three-dimensional (3D) suspension

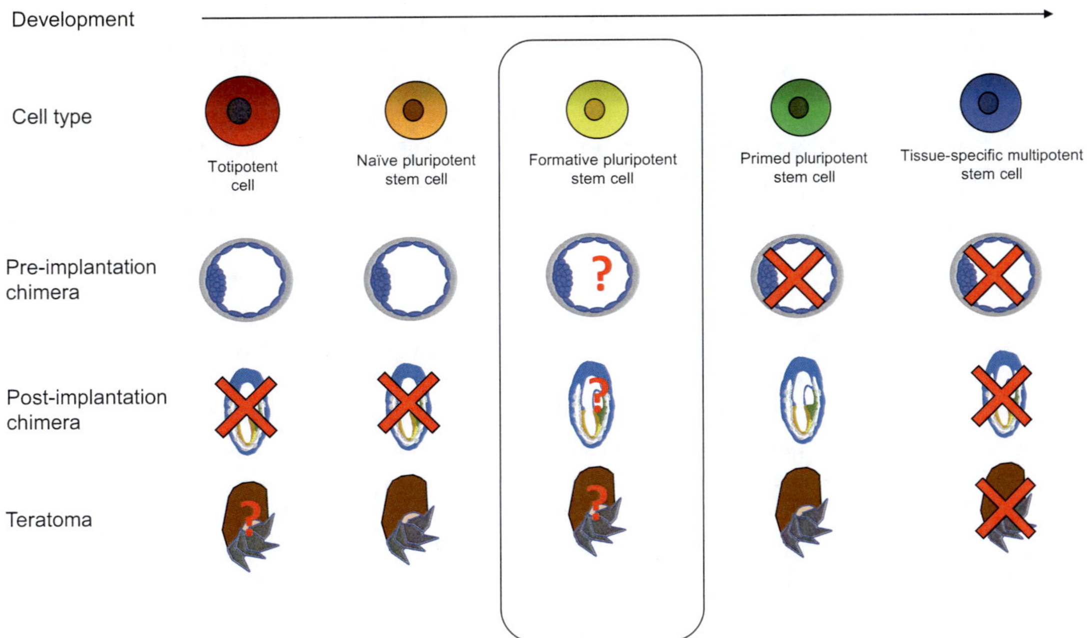

Fig. 1 Stem cell potency. Two main in vivo assays for assessing PS cell potential are chimera formation and teratoma formation. For PS cells, two types of chimera assays can be formed: blastocyst chimerism or epiblast chimerism. Data from these assays assists one in identifying totipotent, naïve pluripotent, formative pluripotent, primed pluripotent, or non-pluripotent cell types. Totipotent cells exhibit autonomous developmental potential and are able to generate an entire organism. Naïve pluripotent stem cells can form teratomas and chimeric animals after injection into preimplantation embryos, but are not considered able to chimerize a postimplantation epiblast. Primed pluripotent stem cells can also form teratomas but cannot generate chimeric animals after injection into preimplantation embryos. They are, however, able to chimerize a postimplantation epiblast. The properties of formative cells are unknown, but they are likely to be teratoma competent and able to chimerize both pre- and postimplantation embryos

culture. Finally, co-culture of PS cells with stromal cells or conditioned media from stromal cell culture can also trigger differentiation. Presently, most attention has focused on the adherent-based and EB-based methodologies for directed differentiation because such experiments are conducted in defined culture conditions where specific factor combinations can be added, enabling greater control to alter experimental conditions when compared to stromal cell-based methods. Exit from pluripotency is indicated by downregulation of pluripotency-associated markers OCT4, SOX2, and NANOG and the concomitant acquisition of lineage-affiliated marker expression. Morphological changes will often follow.

Another assay for evaluating differentiation capacity of test PS cells is teratoma generation after injection of test cells into immunocompromised mice. Teratomas are encapsulated, cystic tumors comprised of all three germ layers. A more detailed analysis of a teratoma can be obtained by requesting that a board-certified pathologist evaluate the histopathological characteristics of a teratoma through analyses of hematoxylin and eosin (H&E)-stained slides. Such an analysis yields more insight into the range of tissues that are formed from each germ layer, the quality of differentiation, the presence of residual undifferentiated cells, the nature of an inflammatory response to teratoma formation, and/or other histopathological features [12].

Although teratoma formation has classically been considered a definitive assay for pluripotency, performance in the teratoma assay may alone not be sufficient to establish the existence of bona fide pluripotency in test cells. For example, some incompletely reprogrammed cells can generate tumors without terminal differentiation [13], while the first iPS cells, which were incompletely reprogrammed, were able to form teratomas [6]. Alternatively, failure to generate teratomas following injection of test PS cells into immunocompromised mice may not necessarily reflect a lack of pluripotency. Some PS cell lines that express high levels of pluripotency-associated markers OCT4, SOX2, and NANOG are able to differentiate in vitro but cannot form teratomas, at least according to currently available evidence [14, 15]. Taken together, these data suggest nuance to interpreting the teratoma assay.

When analyzing data generated from injecting PS cells into immunocompromised mice, it is important to consider experimental artifacts. Although teratomas are commonly injected with Matrigel, co-injection with other matrices or scaffolds can provoke phenomena that can be misidentified as tissue differentiation. In this regard, conducting these types of experiments with proper controls (such as injection of matrix or scaffold alone) and more detailed analysis conducted by a pathologist will aid in determining whether differentiation into the three germ layers occurred. In instances where the teratoma-forming capacity of test cells may be unclear, it may be essential to label donor cells with fluorescent proteins such as GFP to allow one to distinguish donor cell origin from host origin [16]. The use of labeled test cells also excludes an endogenous origin for phenomena that are generated after injecting PS cells into immunocompromised mice. Since teratomas are generated from injecting bulk cultures into immunocompromised mice, another important caveat when interpreting the teratoma assay is that while teratoma formation can help identify cell lines that are pluripotent, it does not evaluate test cell potency at the level of individual cells.

Blastocyst chimera formation assesses the potential of candidate PS cells to reenter development following introduction into a preimplantation embryo. Blastocyst chimeras are generated by aggregating test PS cells with cleavage-stage embryos or injecting test PS cells into eight-cell embryos or blastocysts [17]. Authentic PS cells will colonize all tissues including germ cells. In contrast, compromised PS cells will produce chimeras with low levels of donor cell-derived contributions, retarded development, and/or reduced viability. Compared with the teratoma assay that measures the capacity of a PS cell culture to form tumors, the blastocyst chimera assay is considered a more compelling evidence for pluripotency because it demonstrates functional equivalence between injected test cells and pluripotent cells in the preimplantation embryo.

It is important to note that the inability to form a chimera following injection of test PS cells into preimplantation embryos may not necessarily reflect a lack of pluripotency. As described later, not all PS cell types form chimeras after PS cells are injected into preimplantation embryos. An alternative method for assessing stem cell potential is to graft test PS cells into postimplantation epiblasts in whole-embryo culture and to evaluate contribution to the three germ layers. The ability of PS cell to engraft postimplantation epiblast embryos suggests functional equivalence of test PS cells to pluripotent cells in the postimplantation embryo. In this assay, PS cell clusters are grafted into embryonic epiblasts and their ability to survive, proliferate, and disperse from original graft site is assessed. PS cell clusters may also be grafted into different defined regions of gastrulation-stage embryos: mid-anterior, distal, mid-posterior, and proximal-posterior. Differentiation potential to the three germ layers and germ cells is evaluated. Therefore, while the capacity to form a chimera demonstrates bona fide pluripotency, alternative explanations may account for test cell failure to generate chimeric animals.

Although characterized by higher failure rates, the most rigorous assays assess for pluripotency by germline transmission, tetraploid complementation, or chimera generation from single cells. Germline transmission involves mating chimeras to produce all donor cell-derived offspring, thereby demonstrating the capacity of donor PS cells to give rise to functional gametes. Tetraploid complementation measures the developmental potential of test stem cells to generate a whole organism without the assistance of host pluripotent cells. In this assay, test donor PS cells are injected into tetraploid host embryos. Because tetraploid blastocysts cannot develop normally [17, 18], any resulting embryos will be derived almost entirely from donor test cells. Successful generation of an organism following injection of test cells into tetraploid embryos allows one to conclude that the resulting embryos were generated without nonautonomous rescue of donor cell inadequacies by host cells that might occur during chimera formation.

The assays described above generally refer to assessing the potential of test stem cells at the level of multiple cells. In principle, the potential of a single cell to give rise to all tissue types is the most stringent test for cell potency. Measuring cellular potential at a single-cell level has generally been accomplished by injecting single cells into preimplantation embryos and evaluating chimeric contribution to all tissues of the embryo proper. Germline contribution following introduction of single cells into preimplantation embryos would signify high levels of cellular potency [19]. More recently, single PS cells were shown to generate entire mice by tetraploid complementation, although more studies are needed [20]. It should be noted that these demonstrations of single-cell potency still fall short of the autonomous developmental potential exhibited by totipotent cells in the embryo.

2.2 Key Molecular Features

PS cells are regulated by mechanisms that sustain self-renewal, suppress differentiation, and maintain potential for activation lineage-affiliated regulator expression upon exit from pluripotency. In this section, I provide a basic review of these mechanisms. I emphasize that, despite our ever-evolving understanding of molecular mechanisms governing pluripotency, the central molecular cornerstone of all PS cells remains prominent expression of, as well as dependence on, pluripotent master regulators OCT4 and SOX2.

Some transcription factors (TFs) may be designated as core TFs because they are expressed in all PS cell types. These TFs include OCT4, SOX2, NANOG, and SALL4. NANOG and SALL4 are needed for robust self-renewal, but are not required for PS cell maintenance [21]. Unlike other core TFs, OCT4 and SOX2 (OS) are absolutely required for pluripotency. OCT4 is considered indispensable for pluripotency because loss-of-function studies have demonstrated that OCT4 loss is associated with compromised formation of the pluripotent inner cell mass and loss of the ES cell phenotype, while ectopic induction of OCT4 expression is required for reprogramming of somatic cells [6, 22–26]. OCT4 forms a heterodimer with SOX2. Nearly all cocktails for reprogramming of somatic cells require ectopic expression of OCT4 and SOX2, elevating OS to the pinnacle of pluripotency. Although exogenous OCT4 can be replaced in certain reprogramming cocktails such as when NR5A1 and NR5A2 are overexpressed, PS cells lacking OCT4 have never been described. Therefore, OCT4 is the most critical regulator of pluripotency in all PS cell types [24]. Claims of novel PS cells that lack prominent expression of OCT4 and SOX2 warrant close attention.

It follows that genes identified as essential for pluripotency would also account for the dual hallmarks of PS cells—self-renewal and differentiation potential to the three germ layers—through specialized gene regulatory mechanisms. Studies identifying OS targets in the genome show that OS binds to its own promoters and activates its own expression [23, 24]. OS also bind to the promoters of genes encoding lineage-affiliated TFs [23, 24, 27, 28]. These localization studies support a model in which OS positively regulates itself and other self-renewal genes while suppressing lineage-affiliated TFs. The ability of PS cells to maintain a potential for future activation of germ layer regulator expression may be explained by special types of epigenetic modifications at the promoters of lineage-affiliated TFs. The promoters of lineage-affiliated TFs harbor "bivalent domains" that are decorated with active H3K4me3 and repressive H3K27me3 histone marks [29, 30]. The bivalent configuration is thought to prevent precocious activation of lineage-affiliated regulators while maintaining future potential for induction upon exit from pluripotency [29]. While bivalency may serve as one important mechanism for maintaining differentiation potential, bivalency may not represent the only means by which expression of lineage-affiliated genes is controlled. Precocious transcription of lineage-affiliated genes in some PS cells may be regulated by promoter-proximal pausing of RNA polymerase II rather than the presence of repressive histone modifications such as H3K27me3 or bivalent domains [31].

Despite subtle nuances among the different PS cell types, prominent expression and most compellingly dependence on OCT4 and SOX2 are the strongest discriminating molecular markers for pluripotent status.

3 The Continuum of Murine Pluripotency

3.1 Naïve Pluripotency

Pluripotency comprises a spectrum of different cellular states (Figs. 1 and 2). The laboratory conditions under which PS cells are propagated dictate PS cell state. PS cells established from the inner cell mass (ICM) are called ES cells, whose cardinal attributes are self-renewal, tri-lineage differentiation potential, and most importantly their ability to form chimeras when introduced into preimplantation-stage embryos. When PS cells are generated from developmentally advanced epiblasts, they are called as epiblast stem (EpiS) cells [4, 5]. EpiS cells rarely form chimeras when introduced into preimplantation-stage embryos, but can form chimeric embryos when grafted into postimplantation-stage embryos. Given the functional differences between ES cells and EpiS cells, the terms "naïve" and "primed" were introduced to describe ES cells, EpiS cells, and the characteristics that relate these cell types to the preimplantation and postimplantation pluripotent compartments (Fig. 2) [35]. The ability of a genetically unmodified cell type to participate in high-grade intraspecies chimera formation with germline contribution upon transplantation into a preimplantation embryo is the single most definitive criterion for designation of naïve pluripotency. In other words, while the exact constellation of molecular features that identify a stem cell as naïve or primed differ across studies, the ability to give rise to germline chimeras signifies functional equivalence to the preimplantation epiblast and is therefore considered the quintessential property of naïve pluripotent PS cells. Nonetheless, the distinct pluripotent states are

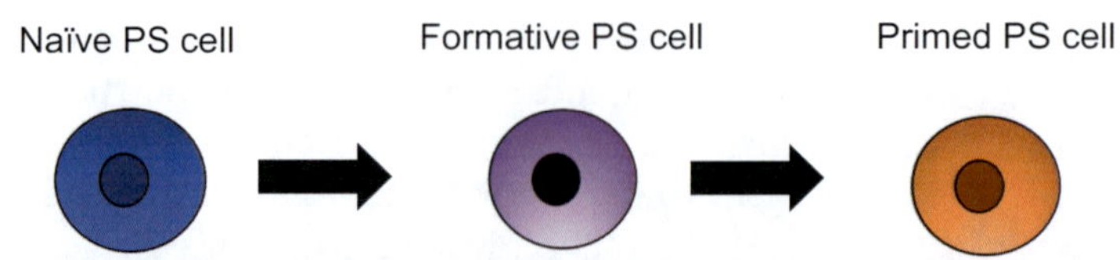

	Naïve	Formative	Primed
TF profile	Naïve TFs Klf4, Tfcp2l1, Klf2 (mouse), KLF17 (human)	Early post-implantation TFs Mouse: Oct6, Otx2? Human: ?	Early post-implantation TFs; Non-ectoderm lineage TFs Foxa2, Brachyury/T
Signaling Requirements	Serum-Free, LIF, ERK inhibition	Serum-Free, WNT inhibition, low ACTIVIN A?,	Serum products, FGF2 and high TGFB/ACTIVIN A
DNA methylation dependence	Tolerate DNMT1 depletion	?	Cannot tolerate DNMT1 depletion
X chromosome status	X reactivation	?	X inactivation
Oct4 enhancer utilization	Oct4 distal enhancer predominance	?	Oct4 proximal enhancer predominance
In vitro germ cell potential	Germ cell incompetence	Germ cell competence	Germ cell incompetence
Pre-implantation chimeras	Yes	Yes?	No
Post-implantation chimeras	No	Yes?	Yes
Examples	2i ESC, a2i ESC, LIF/serum ESC	EpiLC, Rex1- ESC Formative stem cells (ADLA unpublished)	EpiSC, region-selective EpiSC

Fig. 2 Pluripotent-state transitions. Naïve PS cells are characterized by expression of naïve transcription factors (TFs) Klf4, Tfcp2l1 and Klf2 (rodents), or KLF17 (primates). Naïve PS cells in 2i exhibit global DNA hypomethylation reminiscent of the preimplantation blastocyst, whereas naïve PS cells in other conditions exhibit pronounced levels of DNA methylation. Naïve PS cells appear unique in tolerating DNMT1 depletion [32, 33]. Naïve PS cells can form chimeras when introduced into preimplantation embryos. Primed PS cells lack naïve TF expression, manifest high levels of DNA methylation, and cannot form chimeras following introduction into preimplantation blastocyst. However, they can form chimeras when grafted into postimplantation embryos. Formative cells express early postimplantation TFs such as Oct6 and can be identified, at least in rodents, by in vitro germ cell potential [34]. The chimera-forming potential of formative cells has not been examined. The signaling requirements for these different PS cell states differ. Naïve PS cells are propagated in serum-free medium supplemented with LIF and MEK inhibitor. Primed PS cells are propagated in FGF and serum/replacement or high concentrations of ACTIVIN A. Formative PS cells are propagated in serum-free medium supplemented with WNT or Tankyrase inhibitor (ADLA, data unpublished)

governed by different molecular characteristics that can be used to discriminate between different PS cell states.

Naïve and primed PS cells exhibit different molecular features that are thought to reflect similarity to either preimplantation-stage or postimplantation-stage pluripotent epiblast. Such distinguishing molecular features remain relevant because they may be used as surrogate assays to classify cell state. As described above, all PS cells express the master regulators OCT4 and SOX2. At one polar extreme is naïve pluripotency. In mice, naïve cells can be identified when the following four properties are observed: (1) elevated expression levels of regulators designated as naïve TFs: Klf4, Klf2, and Tfcp2l1 [5, 36–39]; (2) enhanced use of the Oct4 distal enhancer; (3) global DNA hypomethylation; and (4) two active X chromosomes in female cells. While each of these markers has been reviewed previously [1], our ever-expanding understanding requires revisiting their relevance. We discuss such key markers below.

In mice, the presence of these four key markers used to identify naïve cells is highly correlated with the prominent expression of naïve pluripotency TFs. The TF circuitry that sustains naïve ES cells (i.e., PS cells maintained in 2i/LIF comprised of MEK inhibitor, GSK3 inhibitor, and leukemia inhibitory factor) has been defined and characterized [39]. Naïve ES cells and preimplantation epiblast express a set of TFs downregulated in EpiS cells and postimplantation embryos, namely Klf4, Klf2, and Tfcp2l1, and to a lesser extent Nanog. These naïve TFs have function in ES cell self-renewal and form an interconnected network [39]. The consolidated expression of naïve TFs observed in 2i/LIF compared with serum/LIF is associated with an increased chimera-forming potential [40]. A crucial observation is that overexpression of any of these "naive" TFs in mouse EpiS cells in conjunction with transfer into 2i/LIF resets primed pluripotency to naïve status. In contrast to parental EpiS cells, the reset PS cells reacquire the potential to give rise to chimeras when introduced into preimplantation embryos. These observations functionally link the so-called naïve TFs Klf4, Klf2, and Tfcp2l1 to chimera competence. Equally important, naïve PS cells are dependent on naïve TFs Klf4, Klf2, and Tfcp2l1, although triple knockdown or knockout of Klf2, Klf4, and Kl5 is needed to reveal the Klf dependency of naïve PS cells.

A characteristic feature of naive PS cells is the preferential use of the distal enhancer (DE) over the proximal enhancer (PE) for Oct4 gene transcription [41]. A switch from preferential use of the DE to the PE is also observed during the transition from preimplantation to postimplantation embryonic development [41]. The loss of Oct4 DE activity and gain of Oct4 DE activity are correlated with loss and gain of chimera competency, respectively [41]. The switch from DE to PE utilization coincides with the downregulation of naïve pluripotency TFs, suggesting that enhancer switching likely signifies rewiring of transcription factor circuitry.

Another feature of pluripotency in the preimplantation embryo is markedly reduced levels of DNA methylation, which rapidly increase during the transition to postimplantation pluripotency. Like preimplantation ICM cells, naïve ES cells cultivated in 2i/LIF also manifest global DNA hypomethylation [36, 42, 43]. It is noteworthy that chimera competent ES cells maintained in culture conditions containing serum maintain methylation levels similarly to postimplantation embryos and somatic tissues [36, 42,

43]. Although less well characterized, chimera-competent ES cells maintained in an alternative 2i comprised of Src inhibitor and GSK3 inhibitor have also been reported to exhibit pronounced DNA methylation levels as well, although more independent studies are needed [44, 45]. Thus, while low levels of DNA methylation reflect similarity to the inner cell mass in one regard, not all chimera-competent PS cells exhibit global DNA hypomethylation. Nonetheless, both 2i/LIF ES cells and serum/LIF ES cells tolerate the loss of DNA methyltransferases, suggesting that naïve PS cells, unlike primed and somatic cells, reside in cellular states that do not require DNA methylation for self-renewal [32, 33, 46–48].

Global DNA hypomethylation may also be associated with the loss of DNA imprints and compromised developmental potential [36, 44]. Moreover, while global hypomethylation resembles the preimplantation epiblast, the genome-wide localization of DNA methylation in naïve PS cells differs from the preimplantation epiblast. Such observations add nuance to interpreting the relevance of global DNA methylation levels for designation of naïve pluripotency. In the face of such molecular complexity, it is important to emphasize that the ability to form chimeras upon introduction into preimplantation embryos remains the quintessential criterion for naïve pluripotency status.

Finally, another feature associated with the primitive naïve condition is two active X chromosomes in female cells [49]. It is noteworthy that pluripotent cells in the preimplantation epiblast, diapaused embryos, and cultured female PS cells exhibit two active X chromosomes. These cells undergo random inactivation when prompted to differentiate [49]. The link between naïve identity and reactivated X chromosome in female cells remains an area of active investigation.

3.2 Primed Pluripotency

At the other polar extreme of the pluripotency spectrum is the final phase of pluripotency, termed primed pluripotency. Primed pluripotency is embodied by mouse EpiS cells. Derived from early postimplantation-stage embryos, EpiS cells, unlike ES cells, inefficiently colonize preimplantation-stage embryos unless they have been genetically modified [5]. In contrast, EpiS cells show chimeric contributions when introduced into cultured postimplantation mouse embryos, supporting the notion that EpiS cells are functionally equivalent to a postimplantation epiblast-like state [50]. At a molecular level, the gene expression signature of EpiS cells is distinct from both naïve preimplantation epiblast and ES cells and instead more closely resembles the anterior primitive streak cells of E7.0 gastrulating epiblasts [51].

Mouse primed EpiS cells exhibit radically distinct characteristics from mouse naïve ES cells with regard to molecular features [35]. Primed EpiS cells lack expression of naïve TFs such as Klf2, Klf4, and Tfcp2l1; show enhanced usage of the Oct4 proximal enhancer; manifest somatic levels of DNA methylation; and show X chromosome inactivation in female cells. Instead, mouse EpiS cells show variable expression of lineage-specific marker genes such as T, Sox1, and Sox17. Propagation of EpiS cells in the laboratory relies on exogenous stimulation of the FGF/ERK cascade. This opposes the MEK inhibitor-containing 2i system, which is employed for maintenance of mouse naïve ES cells.

Less is known about the TF governance of primed PS cells but two features are apparent. First, as described above, primed PS cells lack the expression of naïve TFs such as Klf4, Klf2, and Tfcp2l1. Second, primed EpiS cells express early germ layer-associated markers such as Foxa2 and Brachyury/T. How this transcription factor configuration is related to primed identity is an area of current investigation.

Different culture conditions modulate the properties of primed EpiS cells. Region-selective EpiS cells cultivated in the presence of Tankyrase inhibitors are primed based on the criteria delineated above, but preferentially engraft into posterior epiblasts [50], indicating that the spatial identity of EpiS cells can be shifted by Tankyrase inhibition. These data indicate that our means of classifying PS cell state continuously requires further refinement.

3.3 Alternative and Formative Pluripotent States

While two distinct pluripotent states, termed naïve and primed, are generally considered, various studies have reported alternative mouse PS cell states with putatively "intermediate" identities. These cell types include FAC cells, which are propagated in FGF, ACTIVIN, and CHIR99021 and possess chimera competency and germline potential [52, 53]. However, FAC cells manifest co-expression of naïve TFs and endoderm markers, a pattern of marker expression that is not present in the embryo. This may suggest that "intermediate" cultures, such as FAC cells, may actually comprise a mixture of ES cells and lineage-specified cells or represent an "artifactual" in vitro cell type. Additionally, another putative intermediate PS cell, early primitive ectoderm-like (EPL), has been described [54]. EPL cells do not form chimeras and their signaling requirements remain ill defined. EPL cultures are cultured in the presence of LIF/serum or the amino acid L-proline. The nature of these "intermediate" states of pluripotency is less firmly established, but it will be interesting to determine the relatedness of these alternative PS cell states to pluripotency in the developing embryo.

Studies examining exit from naïve pluripotency provide experimental evidence that the transition from naïve pluripotency may follow a path of phased progression rather than "jumping" into primed-phase pluripotency. Consistent with this notion, Smith and colleagues have proposed a third phase, termed formative pluripotency, which bridges the naïve and primed phases [34]. The formative phase corresponds to the peri-implantation or pregastrula epiblast, a stage in the developing embryo when naïve cells gain competence for lineage specification and primordial germ cell (PGC) fate while also maintaining chimera-forming potential.

An important question is how formative cells can be identified in vitro. In vitro germ cell differentiation potential may distinguish formative stem (FS) cells from naïve and primed-phase cells [34]. Formative-phase cells should also possess a potential to chimerize both preimplantation- and postimplantation-stage embryos [34]. In particular, the ability to efficiently chimerize preimplantation embryos will functionally distinguish formative cells from primed cells, as embryonic pregastrula epiblast also maintains chimera-forming potential, unlike gastrulating epiblast [55]. Formative cells will also possess unique molecular characteristics [56]. Like the naïve and primed PS cell types, formative cells are expected to be governed by a unique transcription factor circuitry that reflects the peri-implantation epiblast [34].

Various populations of formative cells have been reported [56–58]. These cell populations transiently emerge when exit from naïve pluripotency is triggered by withdrawal of 2i/LIF and/or addition of FGF and ACTIVIN A. These populations include epiblast-like cells (EpiL cells) and Rex1::GFP-low ES cells. Unlike naïve and primed PS cells, these transient populations are competent for primordial germ cell (PGC) fate in vitro [56–58]. Molecular analyses of these cells have also revealed different features [56, 59].

It is currently of significant interest to capture stable and homogeneous cultures of PS cells in the formative phase. It should be possible to derive formative cells from naïve PS cells, primed PS cells, and early postimplantation epiblast. Trapping murine cells in a formative phase is likely to be dependent on WNT inhibition and serum deprivation. As delineated above, hypothetical formative-stage PS cells are expected to possess the capacity for primordial germ cell (PGC) fate in vitro, a potential to chimerize both pre- and postimplantation embryos, and distinct transcription factor governance [34]. The capacity to chimerize preimplantation embryos will distinguish formative cells from primed cells. As in vitro representatives of the peri-implantation epiblast, formative cells may possess "superior" differentiation capabilities, including more efficient differentiation to different cell fates and a more unbiased potential to differentiate into the three germ layers. At a transcriptome-wide level, formative cells should form their own cluster distinct from naïve and primed PS cells. Because formative-stage cells constitute a PS cell state distinct from naïve-type cells, they are expected to be unable to self-renew in naïve pluripotency promoting condition 2i/LIF and therefore will lack the expression of naïve TFs Klf4, Klf2, and Tfcp2l1. They will also lack expression of gastrulation-associated TFs such as Brachyury/T and Foxa2. The TFs associated with a peri-implantation identity are still being defined, but POU transcription factor Oct6/Pou3f1 may be a discriminating marker for formative pluripotency. Identifying the TFs that govern formative pluripotency will be essential for identifying formative pluripotency in culture.

4 Considering Human Pluripotent Stem Cells

Conventional human PS cells, whether derived from human embryos or by somatic cell reprogramming, exhibit defining features that align them more closely with EpiS cells than naïve ES cells. It is not possible to evaluate the potential of human PS cells to give rise to intraspecies chimeras due to ethical considerations. Nonetheless, studies have shown that conventional nonhuman primate (NHP) ES cells fail to colonize preimplantation embryos, a property that is reminiscent of EpiS cells. The similarities between NHP ES cells and human ES cells suggest that conventional human PS cells also do not correspond to a chimera-competent state [60]. Because of the lack of human chimeras, molecular comparisons of cultured PS cell lines to cells in human embryos have served as surrogate assays for classification of human PS cell state. Our ever-evolving understanding suggests that it is imperative revisiting these molecular surrogate assays, which I will proceed to do below.

Conventional human PS cells exhibit molecular characteristics similarly manifest in a primed pluripotent state observed in murine EpiS cells, which include lack of naïve TF expression, preferential utilization of the OCT4 proximal enhancer, and elevated levels of DNA methylation [33, 61]. Trending with the murine paradigm, a propensity for X chromosome inactivation has also been observed in female human PS cell lines [61]. However, interpreting X chromosome data in human pluripotency is contentious. I review our ever-changing understanding of these molecular markers below.

An important question is whether molecular hallmarks associated with murine naïve pluripotency are observed during human development. Evidence indicates that this may be the case. For example, marked decreases in DNA methylation levels are observed in both preimplantation and germline development in humans [62–64]. Like mouse preimplantation embryos, human preimplantation embryos also possess a hypomethylated genome. Mouse preimplantation embryos undergo a dramatic re-methylation of the genome upon implantation. If one accepts primed ES cell derivation from preimplantation embryos as a proxy for transition to a postimplantation state, it is noteworthy that transfer of human ICM cells to conventional human ES cell derivation conditions is also accompanied by global genomic re-methylation [63]. Both early-passage and established human PS cells maintain global DNA hypermethylation like adult somatic cells and mouse primed PS cells [62, 63]. The conservation of DNA methylation dynamics during human development suggests that a human naïve cell would be expected to possess a hypomethylated genome, like mouse naïve cells and human preimplantation embryos.

Resetting of DNA methylation levels is thought to be regulated by unique TF circuitry present in mouse naïve PS cells, human preimplantation embryos, and human germline [64]. TFs implicated in regulation of the murine naïve pluripotent state—KLF4 and TFCP2L1—have been detected in human/nonhuman primate preimplantation embryos and human germline and are silenced or exhibit low expression levels in conventional human ES cells [65, 66]. However, subtle but important differences have been observed between rodent and primate ICM gene expression profiles, suggesting some divergence in naïve TF circuitry between rodents and primates. For example, rodent-specific naïve TFs Klf2, Fbxo15, Nr0b1, and Gbx2 are not detected in nonhuman primate marmoset ICM [67]. Conversely, KLF17 is a Kruppel-like factor TF enriched in primate ICM that is not present in mouse ICM but may be a discriminating marker for the human naïve state [67]. These data suggest broad conservation of the gene regulatory network governing naïve pluripotency across rodents and primates with some species-specific differences. Altogether, analyses of gene expression patterns in in vivo compartments known to have naïve pluripotency support the prediction that human PS cells in a putative naïve state will be governed by a similar transcription factor circuitry as that which regulates mouse naïve PS cells [1, 14, 15, 67].

A number of studies have adjusted the state of conventional human PS cells to alternative states that exhibit naïve pluripotency-associated characteristics, such as growth in the presence of MEK and GSK3 inhibitors, amenability to single-cell passaging, and domed morphology [68–72]. Hanna and colleagues described the first transgene-independent naïve-like human PS cells by growing human PS cells in FGF, LIF, TGF-BETA, MEK inhibitor, GSK3 inhibitor, p38 MAPK inhibitor, and JNK inhibitor with optional addition of PKC inhibitor and ROCK inhibitor [68]. Compellingly, the authors showed cross-species chimerism of mouse embryos with human PS cells for the first time, arguing for a functional equivalence with preimplantation epiblast [68]. However, these data have been controversial.

Instead, groups have been evaluating whether candidate naïve cells possess molecular hallmarks associated with murine naïve pluripotency. Generating human PS cells satisfying such criteria was achieved by stabilizing the self-renewal of human PS cells in serum-free N2B27 basal medium supplemented with human LIF and MEK inhibitor [14, 15, 73, 74]. To achieve stability, BRAF or PKC inhibitors were added to LIF and MEK inhibitor-containing media. Intriguingly loss of function for PKC iota and Raf has been shown to inhibit exit from 2i pluripotency in mice. In the first study, N2B27 medium supplemented with a cocktail consisting of LIF, ACTIVIN A, MEK inhibitor, GSK3 inhibitor, BRAF inhibitor, SRC inhibitor, and Y-27632 (5i L/A) generates human naïve PS cells from primed PS cells, human naïve iPS cells, and direct isolation of human naïve ES cells from human blastocysts. A few studies have verified that 5i-based media generates naïve PS cells, although these groups use experimental procedures containing exogenous FGF2. In the second round of studies, KLF2 and NANOG are overexpressed in conventional human PS cells, followed by cultivation in N2B27 medium supplemented with 2iL + PKC inhibitor Go6983 (t2iL + Go conditions). Conventional human PS cells can also be epigenetically reset via treatment with HDAC inhibitor and transfer into t2iL + Go. Finally, dissociated human ICM cells cultured in t2iL + Go supplemented with ROCK inhibitor Y-27632 give rise to human naïve ES cells.

These human naïve PS cells are deemed naïve based on the acquisition of molecular characteristics specific to murine naïve cells, namely, global DNA hypomethylation and expression of TFs implicated in governance of mouse naïve ESCs, such as KLF4 and TFCP2L1 [14, 15]. With regard to DNA methylation, although human naïve PS cells exhibit global DNA hypomethylation, the ability of reported human naïve PS cells to tolerate DNMT1 depletion has not been examined. With regard to naïve TFs, Smith and colleagues showed that human naïve PS cells reset with KLF2 and NANOG collapsed upon depletion of naïve TFs KLF4 and TFCP2L1. Because loss of KLF4 is not sufficient to elicit compromise to mouse naïve PS cell self-renewal, loss of self-renewal in human naïve PS cells upon depletion of a single KLF suggests divergence from the murine paradigm. Some putative human naïve PS cells express KLF17 [67]. While more conclusive analyses of KLF function need to be conducted in transgene-independent human naïve PS cells, dependence on naïve TF KLF circuitry should be considered a stringent marker for human naïve pluripotency [61].

An important caveat to the TF-centric conception of naïve pluripotency is that we do not yet understand the function of naïve TFs in the human embryo. It will be instructive to examine whether KLF4/KLF17 knockout or TFCP2L1 knockout affects cell fate in human embryos.

Because of differences between human and mice, an alternative benchmark for identifying human naïve pluripotency is to directly compare human PS cells to the human preimplantation epiblast. In this regard, it is worth revisiting the relevance of OCT4 distal enhancer utilization, a characteristic of murine naïve cells. The relevance of OCT4 enhancer switching is less firmly established in human cells. Increased utilization of the OCT4 distal enhancer in some human naïve PS cells is apparent. It should be noted that at least one study has described variable OCT4 distal enhancer activity observed in human primed PS cells when using a transgenic reporter containing a sequence from the mouse OCT4 distal enhancer region [75]. Nonetheless, human ES cells harboring a GFP reporter gene knocked into the OCT4 locus containing a deletion of the proximal enhancer (OCT4-deltaPE-GFP) indicate that the OCT4 distal enhancer is not utilized in the primed state. Supporting the importance of this marker, the OCT4-deltaPE-GFP reporter allele also facilitated the design of culture conditions that capture a human PS cell state in which ground-state TFs KLF4 and TFCP2L1 are upregulated [14]. Perhaps the bigger question is whether human embryos exhibit enhanced utilization of the OCT4 distal enhancer. Finally, adding additional complexity, discovery of a previously unidentified human naïve-specific intronic enhancer of OCT4 not present in mouse naïve PS cells highlights human-specific complexity [76]. Therefore, while OCT4 enhancer switching is likely to reflect rewiring of TF circuitry in human cells, it has not yet been possible to relate such changes in enhancer use to gene regulation in the human embryo. An alternative method for assessing similarity to human preimplantation epiblast is to compare the transposon expression of putative human naïve ES cells to those expressed at different stages of human embryo development. However, it is still unclear how changes in transposon expression are related to rewiring of TF circuitry [77].

The significance of observed species-specific differences in X chromosome regulation is contentious. The nature of X chromosome dosage compensation and timing of X chromosome inactivation in human embryos are still being clarified [78, 79]. Additionally, primed human ES cells undergo an epigenetic erosion of the X chromosome that affects their ability to undergo X inactivation [80, 81]. Currently available human naïve PS cells exhibit a shift toward an epiblast-like X chromosome regulation with biallelic expression of X chromosome genes and expression of XIST from one X chromosome [77, 82]. However, such naïve-like PS cells do not appear to undergo random X chromosome inactivation upon exit from pluripotency [77]. In contrast, blastocyst cells are able to undergo random X chromosome inactivation [78]. It will be interesting to resolve these controversies surrounding the X chromosome in different states of human pluripotency. Until then, it remains unclear how X chromosome status will inform classification of PS cell state.

Like naïve mouse PS cells, human naïve PS cells undergo imprint erasure [83]. Such imprint loss is troublesome because imprint abnormalities are associated with cancer. Whether it is possible to uncouple imprint loss from reduced DNA de novo methyltransferase function in human naïve PS cells remains to be demonstrated. For applications of human PS cells in biomedical research and regenerative medicine, an alternative plan may be to reset human PS cells and propagate them for a short period in naïve culture conditions before imprints are lost. Loss of imprints greatly reduces the utility of human naïve PS cells.

In summary, the four markers that identify rodent naïve PS cells—naïve TF expression, Oct4 distal enhancer utilization, global DNA hypomethylation, and X chromosome reactivation—appear to be conserved in human naïve PS cells. However, there exists complexity and nuance. It has not yet been possible to identify which OCT4 enhancer(s) are utilized in the human embryo and the status of the X chromosome in human pluripotency is contentious. It is arguable that, for the time being, rewiring of TF circuitry, delineated by expression levels and dependence on the KLF/TFCP2L1 TFs, and global DNA hypomethylation or tolerance of DNMT1 depletion may be sufficient to identify human naïve pluripotency.

The existence of a rodent formative phase raises the question of a human analog [34, 84]. The human formative phase is likely to differ from the rodent paradigm because primate and human embryos form an embryonic disk rather than the egg cylinder observed in rodents and because primate epiblast development is protracted relative to rodent epiblast development [84]. In humans, in utero postimplantation formative epiblast cannot be sampled, so analyses of murine and primate formative pluripotency will have to suffice. For analysis of human formative pluripotency, the appropriate initial tissue type may be an inner cell mass (ICM) explant culture that develops in vitro to a putative formative embryonic disk stage. Moreover, human naïve PS cells may serve as the appropriate starting cell type for derivation of human formative-stage cells in culture, although it should be possible to reprogram somatic cells directly to a formative phase.

It will be interesting to determine whether the various reported alternative PS cell types possess characteristics associated with formative-stage pluripotency. Unfortunately, it remains unclear how to validate cultures of putative human formative-phase cells because the formative phase is still poorly defined. As primed PS cells can generate germ cells, in vitro germ cell potential may not be sufficient to identify a putative human formative cell [85–88]. As in vitro culture representatives of the early postimplantation epiblast, formative cells are anticipated to differentiate more efficiently into the three germ layers, unlike conventional human PS cells that tend to be lineage biased. The TFs that regulate formative-phase pluripotency have not been elucidated, but the POU TF OCT6/POU3F1 is a leading candidate marker for identifying formative status in human cells. Human formative cells will lack the expression of naïve TFs KLF4, TFCP2L1, and KLF17 and also lack the expression of primitive streak markers. Formative stem cells are anticipated to possess intermediate or pronounced levels of DNA methylation that may enable genetic stability and imprint stability.

Defining the nature of human formative pluripotency and capturing such a state in culture are of current and future scientific interest [34, 84].

5 Identifying Chimera-Competent Human Pluripotency

Recently, interest has grown in examining the ability of PS cells from one species to contribute to the embryos of another species [89]. The generation of interspecies chimeras between mice and rats is reproducible [89]. The introduction of human PS cells into embryos of mice or pigs has not given rise to live interspecies chimeras or meaningful levels of human donor cell-derived chimerism [14, 20, 53, 77]. Although the generation of live interspecies chimeras using human cells is ethically restricted, studies that have attempted to generate interspecies chimeras report very limited human contribution to embryos of other species [14, 20, 53, 68, 77]. It is unclear why very low levels of human chimerism have been observed so far, but one possibility is that achieving sufficient matching developmental stage of injected donor cells with host embryos may be needed for high-grade chimera formation [90].

Some evidence suggests that the principles that underlie chimera competency in primates may differ from the rodent paradigm. Unlike mouse ICM cells, primate ICM cells fail to form blastocyst chimeras, suggesting important differences between rodents and primates [60]. In contrast, aggregation of earlier stage blastomeres does produce chimerism in primates [60]. Adding additional complexity, primate PS cells do not appear as refractory to chimera formation as ICM cells. This is reflected in a recent study that described generation of intraspecies chimeric fetuses by injection of naïve-like NHP cells into preimplantation embryos. The resulting fetuses had contribution to all three embryonic germ layers and the germline. An important caveat to these observations is that contributions were of low grade [91].

To enhance interspecies chimerism, efforts have been made to match donor cell stage with host embryos by resetting the developmental stage of primate and human PS cells toward a naïve state. Despite these adjustments in human PS cell characteristics, little success has been achieved. Limited evidence for interspecies chimerism has been described in a study that injected primate naïve-like iPSCs into mouse blastocysts [92]. Attempts to generate interspecies chimeras using human naïve-like cells and either mouse or pig host embryos have also produced very low rates of chimerism [20, 53, 77]. The limited interspecies chimerism achieved with human naïve-like PS cells may reflect the presence of species barriers beyond matching developmental timing.

In spite of attempts to adjust the developmental stage of primed PS cells to a naïve state, the developmental stage of donor cells and host embryos may still have not been sufficiently matched [90]. This could be because naïve pluripotency has not yet been fully achieved. Alternatively, as stated above, the parameters involved in identifying a chimera-competent cellular state in primates may fundamentally differ from the rodent paradigm. It remains unclear whether human reset stem cells analogous to rodent naïve PS cells are able to reenter development efficiently upon introduction into an embryo [1, 93]. In other words, the molecular criteria used to identify naïve pluripotency in rodents, such as naïve TF governance, may not necessarily capture a chimera-competent cellular state in primates [1, 93]. While more experiments are needed, currently available evidence suggests that alternative types of PS cells, rather than naïve-like types, may possess a higher capacity for interspecies chimera generation [53].

Evolutionary distance between donor and host species may also play a role in failure to generate interspecies chimeras. The generation of high-grade chimeras between closely related species—such as mice and rats—suggests that the results obtained when introducing human PS cells into the embryos of more distantly related species, such as sheep and pigs, will differ if host embryos from more closely related species such as nonhuman primates (NHPs) were used [20, 53, 77]. Moreover, the introduction of cultured human PS cells into human preimplantation embryos might comprise a well-controlled system to examine chimera-competency in human cells. To develop approaches to overcome species barriers, it will be instructive to generate human-human or human-monkey chimeric embryos. By studying the chimerism in early-stage human-human human-monkey embryos cultured to postimplantation stages, it may be possible to identify impediments to human-human or human-nonprimate interspecies chimera formation. Such ethically contentious experiments should proceed under legal, social, and ethical guidelines.

6 Discussion

PS cells have revolutionized biology. Pluripotency encompasses a spectrum of different cellular states. Historically, mouse and human PS cells exhibited dramatic differences that were previously ascribed to species differences. The discovery of mouse EpiS cells taught us that many of the different properties of mouse and human PS cells may not originate from species-specific differences but may instead be attributable to relatedness to different developmental phases of pluripotency. Additionally, we learned that PS cell types with radically distinct molecular and biological characteristics can be generated by modulating culture parameters [4, 5]. The identification of two attractor states in mice—naïve and primed—provoked the question of whether it was possible to reset conventional human PS cells, which correspond to a primed state, to a cellular state analogous to that inhabited by mouse embryonic stem cells [35].

The quest for obtaining stable human naïve PS cells in culture ignited a flurry of studies leading to the isolation of many different PS cell states [14, 15, 68]. The spectrum of PS cell states challenges the paradigm of naïve-primed bistability that has dominated our thinking of pluripotency in the last decade [34]. Currently, more attention is being devoted to understanding the phased progression of pluripotency and identifying the signaling requirements for propagation of formative cells.

Recently, some groups have reported the generation of PS cells with potential for both embryonic and extraembryonic fates [20]. The potential to form both embryonic and trophoblastic fates is broadly accepted as one feature that distinguishes totipotency from pluripotency, although totipotent cells exhibit autonomous developmental potential, which recently reported expanded potential stem (EPS) cells do not. In light of the rapid advances being made in the field of pluripotency, particularly regarding artificial embryos and expanded potential stem cells, proper evaluation of novel claims of totipotency remains relevant [20]. These distinct stem cell states are less firmly established in the pluripotency field. This chapter does not address these claims of "totipotent" or "expanded potential" cells and I will refer readers to the previous "Hallmarks" review that outlines how to approach these claims [1].

Despite limited success thus far, scientists are continuing to explore how to generate live chimeras comprised of cells from humans and other species [14, 53, 68, 77]. It remains unclear whether human reset cells with naïve TF governance are indeed chimera competent, or whether and how the principles underlying chimera competence in humans and primates diverge from the rodent paradigm. Resolving this question may require more attempts at generating intraspecies chimeras using candidate non-human primate naïve PS cells or generation of human-monkey chimeras, which is ethically contentious [94]. Studies involving the introduction of cultured human stem cells into human pre-implantation embryos are likely to provide further insights into this question.

In the future, patient-specific PS cells may be used not only for modeling disease, testing drugs, or generating transplantable cells, but also for producing patient-specific organs as well.

References

1. De Los Angeles A, Ferrari F, Xi R, Fujiwara Y, Benvenisty N, Deng H, Hochedlinger K, Jaenisch R, Lee S, Leitch HG, Lensch MW, Lujan E, Pei D, Rossant J, Wernig M, Park PJ, Daley GQ (2015) Hallmarks of pluripotency. Nature 525:469–478
2. Evans MJ, Kaufman MH (1981) Establishment in culture of pluripotential cells from mouse embryos. Nature 292:154–156
3. Martin GR (1981) Isolation of a pluripotent cell line from early mouse embryos cultured in medium conditioned by teratocarcinoma stem cells. Proc Natl Acad Sci U S A 78:7634–7638
4. Brons IG, Smithers LE, Trotter MW, Rugg-Gunn P, Sun B, Chuva de Sousa Lopes SM, Howlett SK, Clarkson A, Ahrlund-Richter L, Pedersen RA, Vallier L (2007) Derivation of pluripotent epiblast stem cells from mammalian embryos. Nature 448:191–195
5. Tesar PJ, Chenoweth JG, Brook FA, Davies TJ, Evans EP, Mack DL, Gardner RL, McKay RD (2007) New cell lines from mouse epiblast share defining features with human embryonic stem cells. Nature 448:196–199
6. Takahashi K, Yamanaka S (2007) Induction of pluripotent stem cells from mouse embryonic and adult fibroblast cultures by defined factors. Cell 126:663–676
7. Thomson JA, Itskovitz-Eldor J, Shapiro SS, Waknitz MA, Swiergel JJ, Marshall VS, Jones JM (1998) Embryonic stem cells derived from human blastocysts. Science 282:1145–1147
8. Thomson JA, Kalishman J, Golos TG, Durning M, Harris CP, Becker RA, Hearn JP (1995) Isolation of a primate embryonic stem cell line. PNAS 92:7844–7848
9. Takahashi K, Tanabe K, Ohnuki M, Narita M, Ichisaka T, Tomoda K, Yamanaka S (2007) Induction of pluripotent stem cells from adult human fibroblasts by defined factors. Cell 131:861–872
10. Dimos JT, Rodolfa KT, Niakan KK, Weisenthal LM, Mitsumoto H, Chung W, Croft GF, Saphier G, Leibel R, Goland R, Wichterle H, Henderson CE, Eggan K (2008) Induced pluripotent stem cells generated from patients with ALS can be differentiated into motor neurons. Science 321:1281–1221
11. Martello G, Smith A (2014) The nature of embryonic stem cells. Annu Rev Cell Dev Biol 30:647–675
12. Damjanov I, Andrews PW (2016) Teratomas produced from human pluripotent stem cells xenografted into immunodeficient mice – a histopathology atlas. Int J Dev Biol 60:337–419
13. Chan EM, Ratanasirintrawoot S, Park IH, Manos PD, Loh YH, Huo H, Miller JD, Hartung O, Rho J, Ince TA, Daley GQ, Schlaeger TM (2009) Live cell imaging distinguishes bona fide human iPS cells from partially reprogrammed cells. Nat Biotechnol 28:1033–1037
14. Theunissen TW, Powell BE, Wang H, Mitalipova M, Faddah DA, Reddy J, Fan ZP, Maetzel D, Ganz K, Shi L, Lungjangwa T, Imsoonthornruksa S, Stelzer Y, Rangarajan S, D'Alessio A, Zhang J, Gao Q, Dawlaty MM, Young RA, Gray NS, Jaenisch R (2014) Systematic identification of culture conditions for induction and maintenance of naïve human pluripotency. Cell Stem Cell 15:471–487
15. Takashima Y, Guo G, Loos R, Nichols J, Ficz G, Krueger F, Oxley D, Santos F, Clarke J, Mansfield W, Reik W, Bertone P, Smith A (2014) Resetting transcription factor control circuitry toward ground-state pluripotency in human. Cell 158:1254–1269
16. De Los Angeles A, Ferrari F, Fujiwara Y, Mathieu R, Lee S, Lee S, Tu HC, Ross S, Chou S, Nguyen M, Wu Z, Theunissen TW, Powell BE, Imsoonthornruksa S, Chen J, Borkent M, Krupalnik V, Lujan E, Wernig M, Hanna JH, Hochedlinger K, Pei D, Jaenisch R, Deng H, Orkin SH, Park PJ, Daley GQ (2015) Failure to replicate the STAP cell phenomenon. Nature 525:E6–E9
17. Nagy A, Rossant J, Nagy R, Abramow-Newerly W, Roder JC (1993) Derivation of completely cell culture-derived mice from early-passage embryonic stem cells. Proc Natl Acad Sci U S A 90:8424–8428
18. Nagy A, Gocza E, Diaz EM, Prideaux VR, Ivanyi E, Markkula M, Rossant J (1990) Embryonic stem cells alone are able to support fetal development in the mouse. Development 110:815–821
19. Wang Z, Jaenisch R (2004) At most three ES cells contribute to the somatic lineages of chimeric mice produced by ES-tetraploid complementation. Dev Biol 275:192–201
20. Yang Y, Liu B, Xu J, Wang J, Wu J, Shi C, Xu Y, Dong J, Wang C, Lai W, Zhu J, Xiong L, Zhu D, Li X, Yang W, Yamauchi T, Sugawara A, Li Z, Sun F, Li X, Li C, He A, Du Y, Wang T, Zhao C, Li H, Chi X, Zhang H, Liu Y, Li C, Duo S, Yin M, Shen H, Belmonte JCI, Deng H (2017) Derivation of pluripotent stem cells with in vivo embryonic and extraembryonic potency. Cell 169:243–257

21. Sakaki-Yumoto M, Kobayashi C, Sato A, Fujimura S, Matsumoto Y, Takasato M, Kodama T, Aburatani H, Asashima M, Yoshida N, Nishinakamura R (2006) The murine homolog of SALL4, a causative gene in Okihiro syndrome, is essential for embryonic stem cell proliferation, and cooperates with SALL1 in anorectal, heart, brain, and kidney development. Development 133:3005–3013
22. Boyer LA, Lee TI, Cole MF, Johnstone SE, Levine SS, Zucker JP, Guenther MG, Kumar RM, Murray HL, Jenner RG, Gifford DK, Melton DA, Jaenisch R, Young RA (2005) Core regulatory circuitry in human embryonic stem cells. Cell 122:947–956
23. Loh YH, Wu Q, Chew JL, Vega VB, Zhang W, Chen X, Bourque G, George J, Leong B, Liu J, Wong KY, Sung KW, Lee CW, Zhao XD, Chiu KP, Lipovich L, Kuznetsov VA, Robson P, Stanton LW, Wei CL, Ruan Y, Lim B, Ng HH (2006) The Oct4 and Nanog transcription network regulates pluripotency in mouse embryonic stem cells. Nat Genet 38(4):431–440
24. Nichols J, Zevnik B, Anastassiadis K, Niwa H, Klewe-Nebenius D, Chambers I, Scholer H, Smith A (1998) Formation of pluripotent stem cells in the mammalian embryo depends on the POU transcription factor Oct4. Cell 95:379–391
25. Chambers I, Colby D, Robertson M, Nichols J, Lee S, Tweedie S, Smith A (2003) Functional expression cloning of Nanog, a pluripotency sustaining factor in embryonic stem cells. Cell 113:643–655
26. Mitsui K, Tokuzawa Y, Itoh H, Segawa K, Murakami M, Takahashi K, Maruyama M, Maeda M, Yamanaka S (2003) The homeoprotein Nanog is required for maintenance of pluripotency in mouse epiblast and ES cells. Cell 113:631–642
27. Chen X, Xu H, Yuan P, Fang F, Huss M, Vega VB, Wong E, Orlov YL, Zhang W, Jiang J, Loh YH, Yeo HC, Yeo ZX, Narang V, Govindarajan KR, Leong B, Shahab A, Ruan Y, Bourque G, Sung WK, Clarke ND, Wei CL, Ng HH (2008) Integration of external signaling pathway with the core transcriptional network in embryonic stem cells. Cell 133:1106–1117
28. Kim J, Chu J, Shen X, Wang J, Orkin SH (2008) An extended transcriptional network for pluripotency of embryonic stem cells. Cell 132:1049–1061
29. Bernstein BE, Mikkelsen TS, Xie X, Kamal M, Huebert DJ, Cuff J, Fry B, Meissner A, Wernig M, Plath K, Jaenisch R, Wagschal A, Feil R, Schreiber SL, Lander ES (2006) A bivalent chromatin structure marks key developmental genes in embryonic stem cells. Cell 125:315–326
30. Mikkelsen TS, Ku M, Jaffe DB, Issac B, Lieberman E, Giannoukos G, Alvarez P, Brockman W, Kim TK, Koche RP, Lee W, Mendenhall E, O'Donovan A, Presser A, Russ C, Xie X, Meissner A, Wernig M, Jaenisch R, Nusbaum C, Lander ES, Bernstein BE (2007) Genome-wide maps of chromatin state in pluripotent and lineage-committed cells. Nature 448:553–560
31. Marks H, Kalkan T, Menafra R, Dennisov S, Jones K, Hofemeister H, Nichols J, Kranz A, Stewart AF, Smith A, Stunnenberg HG (2012) The transcriptional and epigenomic foundations of ground state pluripotency. Cell 149:590–604
32. Geula S, Moshitch-Moshokovitz S, Dominissini D, Mansour AA, Kol N, Salmon-Divon M, Hershkovitz V, Peer E, Mor N, Manor YS, Ben-Haim MS, Eyal E, Yunger S, Pinto Y, Jaitin DA, Viukov S, Rais Y, Krupalnik V, Chomsky E, Zerbib M, Maza I, Rechavi Y, Massarwa R, Hanna S, Amit I, Levanon EY, Amariglio N, Stern-Ginossar N, Novershtern N, Rechavi G, Hanna JH (2015) m6A mRNA methylation facilitates resolution of naïve pluripotency toward differentiation. Science 347:1002–1006
33. Liao J, Karnik R, Gu H, Ziller MJ, Clement K, Tsankov AM, Akopian V, Gifford CA, Donaghey J, Galonska C, Pop R, Reyon D, Tsai SQ, Mallard W, Joung JK, Rinn JL, Gnirke A, Meissner A (2015) Targeted disruption of DNMT1, DNMT3A, and DNMT3B in human embryonic stem cells. Nat Genet 47:469–478
34. Smith A (2017) Formative pluripotency: the executive phase in a developmental continuum. Development 144:365–373
35. Nichols J, Smith A (2009) Naïve and primed pluripotent states. Cell Stem Cell 4:487–492
36. Leitch HG, McEwen KR, Turp A, Encheva V, Carroll T, Grabole N, Mansfield W, Nashun B, Knezovich JG, Smith A, Surani MA, Hajkova P (2013) Naïve pluripotency is associated with global DNA hypomethylation. Nat Struct Mol Biol 20:311–316
37. Guo G, Yang J, Nichols J, Hall JS, Eyres I, Mansfield W, Smith A (2009) Klf4 reverts developmentally programmed restriction of ground state pluripotency. Development 136:1063–1069
38. Bao S, Tang F, Li X, Hayashi K, Gillich A, Lao K, Surani MA (2009) Epigenetic reversion of post-implantation epiblast to pluripotent embryonic stem cells. Nature 461:1292–1295

39. Dunn SJ, Martello G, Yordanov B, Emmott S, Smith AG (2014) Defining an essential transcription factor program for naïve pluripotency. Science 344:1156–1160
40. Alexandrova S, Kalkan T, Humphreys P, Riddell A, Scognamigliio R, Trumpp A, Nichols J (2016) Selection and dynamics of embryonic stem cell integration into early mouse embryos. Development 143:24–34
41. Han DW, Tapia N, Joo JY, Greber B, Arauzo-Bravo MJ, Bernemann C, Ko K, Wu G, Stehling M, Do JT, Scholer HR (2010) Epiblast stem cell subpopulations represent mouse embryos of distinct pregastrulation stages. Cell 143:617–627
42. Ficz G, Hore TA, Santos F, Lee HJ, Dean W, Arand J, Krueger F, Oxley D, Paul YL, Walter J, Cook SJ, Andrews S, Branco MR, Reik W (2013) FGF signaling inhibition in ESCs drives rapid genome-wide demethylation to the epigenetic ground state of pluripotency. Cell Stem Cell 13:351–359
43. Habibi E, Brinkman AB, Arand J, Kroeze LI, Kerstens HH, Matarese F, Lepikhov K, Gut M, Brun-Heath I, Hubner NC, Benedetti R, Altucci L, Janesen JH, Walter J, Gut IG, Marks H, Stunnenberg HG (2013) Whole-genomebisulfite sequencing of two distinct interconvertible DNA methylomes of mouse embryonic stem cells. Cell Stem Cell 13:360–369
44. Choi J, Huebner AJ, Clement K, Walsh RM, Savol A, Lin K, Gu H, DiStefano B, Brumbaugh J, Kim S-Y, Sharif J, Rose CM, Mohammad A, Odajima J, Charron J, Shioda T, Gnirke A, Gygi S, Koseki H, Sadreyev RI, Xiao A, Meissner A, Hochedlinger K (2017) Prolonged Mek1/2 suppression impairs the developmental potential of embryonic stem cells. Nature 548:219–223
45. Yagi M, Kishigami S, Tanaka A, Semi K, Mizutani E, Wakayama S, Wakayama T (2017) Derivation of ground-state female ES cells maintaining gamete-derived DNA methylation. Nature 548:224–227
46. Li E, Bestor TH, Jaenisch R (1992) Target mutation of the DNA methyltransferase gene results in embryonic lethality. Cell 69:915–926
47. Jackson-Grusby L, Beard C, Possemato R, Tudor M, Fambrough D, Csankovski G, Dausman J, Lee P, Wilson C, Lander E, Jaenisch R (2001) Loss of genomic methylation causes p53-dependent apoptosis and epigenetic deregulation. Nat Genet 27:31–39
48. Wernig M, Meissner A, Foreman R, Brambrink T, Ku M, Hochedlinger K, Bernstein BE, Jaenisch R (2007) In vitro reprogramming of fibroblasts into a pluripotent ES-cell-like state. Nature 448:318–324
49. Silva J, Nichols J, Theunissen TW, Guo G, van Oosten AL, Barrandon O, Wray J, Yamanaka S, Chambers I, Smith A (2009) Nanog is the gateway to the pluripotent ground state. Cell 138:722–737
50. Wu J, Okamura D, Li M, Suzuki K, Luo C, Ma L, He Y, Li Z, Benner C, Tamura I, Krause MN, Nery JR, Du T, Zhang Z, Hishida T, Takahashi Y, Aizawa E, Kim NY, Lajara J, Guillen P, Campistol JM, Esteban CR, Ross PJ, Saghatelian A, Ren B, Ecker JR, Izpisua Belmonte JC (2015) An alternative pluripotent state confers interspecies chimeric competency. Nature 521:316–321
51. Kojima Y, Kaufman-Francis K, Studdert JB, Steiner KA, Power MD, Loebel DA, Jones V, Hor A, de Alencastro G, Logan GJ, Teber ET, Tam OH, Stutz MD, Alexander IE, Pickett HA, Tam PP (2014) The transcriptional and functional properties of mouse epiblast stem cells resemble the anterior primitive streak. Cell Stem Cell 14:107–120
52. Tsukiyama T, Ohinata Y (2014) A modified EpiSC culture condition containing a GSK3 inhibitor can support germline-competent pluripotency in mice. PLoS One 9:e95329
53. Wu J, Platero-Luengo A, Sakurai M, Sugawara A, Gil MA, Yamauchi T, Suzuki K, Bogliotti YS, Cuello C, Morales Valencia M, Okumura D, Luo J, Vilarino M, Parrilla I, Soto DA, Martinez CA, Hishida T, Sanchez-Bautista S, Martinez-Martinez ML, Wang H, Nohalez A, Aizawa E, Martinez-Redondo P, Ocampo A, Reddy P, Roca J, Maga EA, Esteban CR, Berggren WT, Nunez Delicado E, Lajara J, Guillen I, Guillen P, Campistol JM, Martinez EA, Ross PJ, Izpisua Belmonte JC (2017) Interspecies chimerism with mammalian pluripotent stem cells. Cell 168:473–486
54. Rathjen J, Lake JA, Bettess MD, Washington JM, Chapman G, Rathjen PD (1999) Formation of a primitive ectoderm like cell population, EPL cells, from ES cells in response to biologically derived factors. J Cell Sci 112:601–612
55. Gardner RL, Lyon MF, Evans EP, Burtenshaw MD (1985) Clonal analysis of X-chromosome inactivation and the origin of the germ line in the mouse embryo. J Embryol Exp Morphol 88:349–363
56. Kalkan T, Olova N, Roode M, Mulas C, Lee HJ, Nett I, Marks H, Walker R, Stunnenberg HG, Lilley KS, Nichols J, Reik W, Bertone P, Smith A (2017) Tracking the embryonic stem cell transition from ground state pluripotency. Development 144:1221–1234

57. Hayashi K, Ohta H, Kurimoto K, Aramaki S, Saitou M (2011) Reconstitution of the mouse germ cell specification pathway in culture by pluripotent stem cells. Cell 146:519–532
58. Mulas C, Kalkan T, Smith A (2017) NODAL secures pluripotency upon embryonic stem cell progression from the ground state. Stem Cell Reports 9:77–91
59. Kurimoto K, Yabuta Y, Hayashi K, Ohta H, Kiyonari H, Mitani T, Moritoki Y, Kohri K, Kimura H, Yamamoto T, Katou Y, Shirahige K, Saitou M (2015) Quantitative dynamics of chromatin remodeling during germ cell specification from mouse embryonic stem cells. Cell Stem Cell 16:517–532
60. Tachibana M, Sparman M, Ramsey C, Ma H, Lee H-S, Penedo MC, Mitalipov S (2012) Generation of chimeric rhesus monkeys. Cell 148:285–295
61. Hanna J, Cheng AW, Saha K, Kim J, Lengner CJ, Soldner F, Cassady JP, Muffat J, Carey BW, Jaenisch R (2010) Human embryonic stem cells with biological and epigenetic characteristics similar to those of mouse ESCs. PNAS 107:9222–9227
62. Guo H, Zhu P, Yan L, Li R, Hu B, Lian Y, Yan J, Ren X, Lin S, Li J, Jin X, Shi X, Liu P, Wang X, Wang W, Wei Y, Li X, Guo F, Wu X, Fan X, Yong J, Wen L, Xie SX, Tang F, Qiao J (2014) The DNA methylation landscape of human early embryos. Nature 511:606–610
63. Smith ZD, Chan MM, Humm KC, Karnik R, Mekhoubad S, Regev A, Eggan K, Meissner A (2014) DNA methylation dynamics of the human preimplantation embryo. Nature 511:611–615
64. Tang WW, Dietmann S, Irie N, Leitch HG, Floros VI, Bradshaw CR, Hackett JA, Chinnery PF, Surani MA (2015) A unique gene regulatory network resets the human germline epigenome for development. Cell 161:1453–1467
65. Yan L, Yang M, Guo H, Yang L, Wu J, Li R, Liu P, Lian Y, Zheng X, Yan J, Huang J, Li M, Wu X, WenL LK, Li R, Qiao J, Tang F (2013) Single-cell RNA-seq profiling of human preimplantation embryos and embryonic stem cells. Nat Struct Mol Biol 20:1131–1139
66. Blakeley P, Fogarty NM, del Valle I, Wamaitha SE, Hu TX, Elder K, Snell P, Christie L, Robson P, Niakan KK (2015) Defining the three cell lineages of the human blastocyst by single-cell RNA seq. Development 142:3151–3165
67. Boroviak T, Loos R, Lombard P, Okahara J, Behr R, Sasaki E, Nichols J, Smith A, Bertone P (2015) Lineage-specific profiling delineates the emergence and progression of naïve pluripotency in mammalian embryogenesis. Dev Cell 35:366–382
68. Gafni O, Weinberger L, Mansour AA, Manor YS, Chomsky E, Ben-Yosef D, Kalma Y, Viukov S, Maza I, Zviran A, Rais Y, Shipony Z, Mukamel Z, Krupalnik V, Zerbib M, Geula S, Caspi I, Schneir D, Shwartz T, Gilad S, Amann-Zalcenstein D, Benjamin S, Amit I, Tanay A, Massarwa R, Novershtern N, Hanna JH (2013) Derivation of novel human ground state naïve pluripotent stem cells. Nature 504:282–286
69. Chan YS, Goke J, Ng JH, Lu X, Gonzalez KA, Tan CP, Tng WQ, Hong ZZ, Lim YS, Ng HH (2013) Induction of a human pluripotent state with distinct regulatory circuitry that resembles preimplantation epiblast. Cell Stem Cell 13:663–675
70. Ware CB, Nelson AM, Mecham B, Hesson J, Zhou W, Jonlin EC, Jimenez-Caliani AJ, Deng X, Cavanaugh C, Cook S, Tesar PJ, Okada J, Margaretha L, Sperber H, Choi M, Blau CA, Treuting PM, Hawkins RD, Cirulli V, Ruohola-Baker H (2014) Derivation of naïve human embryonic stem cells. PNAS 111:4484–4489
71. Valamehr B, Robinson M, Abujarour R, Rezner B, Vranceanu F, Le T, Medcalf A, Lee TT, Fitch M, Robbins D, Flynn P (2014) Platform for induction and maintenance of transgene-free hiPSCs resembling ground state pluripotent stem cells. Stem Cell Reports 2:366–381
72. Duggal G, Warrier S, Ghimire S, Broekaert D, Van der Jeught M, Lierman S, Deroo T, Peelman L, Van Soom A, Cornelissen R, Menten B, Mestdagh P, Vandesompele J, Roost M, Slieker RC, Heijmans BT, Deforce D, De Sutter P, De Sousa Lopes SC, Heindryckx B (2015) Alternative routes to induce naïve pluripotency in human embryonic stem cells. Stem Cells 33:2686–2698
73. Guo G, von Meyenn F, Rostovskaya M, Clarke J, Dietmann S, Baker D, Sahakyan A, Myers S, Bertone P, Reik W, Plath K, Smith A (2017) Epigenetic resetting of human pluripotency. Development 144:2748–2763
74. Guo G, von Meyenn F, Santos F, Chen Y, Reik W, Bertone P, Smith A, Nichols J (2016) Naïve pluripotent stem cells derived directly from isolated cells of the human inner cell mass. Stem Cell Reports 6:437–446
75. Hotta A, Cheung AY, Farra N, Vijayaragavan K, Seguin CA, Draper JS, Pasceri P, Maksakova IA, Mager DL, Rossant J, Bhatia M, Ellis J (2009) Isolation of human iPS cells using EOS lentiviral vectors

to select for pluripotency. Nat Methods 6:370–376
76. Pastor WA, Liu W, Chen D, Ho J, Kim R, Hunt TJ, Lukianchikov A, Liu X, Polo JM, Jacobsen SE, Clark AT (2018) TFAP2C regulates transcription in human naïve pluripotency by opening enhancers. Nat Cell Biol 20:553–564
77. Theunissen TW, Friedli M, He Y, Planet E, O'Neil RC, Markoulaki S, Pontis J, Wang H, Iouranova A, Imbeault M, Duc J, Cohen MA, Wert KJ, Castanon R, Zhang Z, Huang Y, Nery JR, Drotar J, Lungjangwa T, Trono D, Ecker JR, Jaenisch R (2016) Molecular criteria for defining the naïve human pluripotent state. Cell Stem Cell 19:502–515
78. Okamoto I, Patrat C, Thepot D, Peynot N, Fauque P, Daniel N, Diabangouaya P, Wolf JP, Renard JP, Duranthon V, Heard E (2011) Eutherian mammals use diverse strategies to initiate X-chromosome inactivation during development. Nature 472:370–374
79. O'Leary T, Heindryckx B, Lierman S, van Bruggen D, Goeman JJ, Vandewoestyne M, Deforce D, de Sousa Lopes SM, De Sutter P (2012) Tracking the progression of the human inner cell mass during embryonic stem cell derivation. Nat Biotechnol 30:278–282
80. Silva SS, Rowntree RK, Mekhoubad S, Lee JT (2008) X-chromosome inactivation and epigenetic fluidity in human embryonic stem cells. PNAS 105:4820–4825
81. Anguera MC, Sadreyev R, Zhang Z, Szanto A, Payer B, Sheridan SD, Kwok S, Haggarty SJ, Sur M, Alvarez J, Gimelbrant A, Mitalipova M, Kirby JE, Lee JT (2012) Molecular signatures of human induced pluripotent stem cells highlight sex differences and cancer genes. Cell Stem Cell 11:75–90
82. Sahakyan A, Kim R, Chronis C, Sabri S, Bonora G, Theunissen TW, Kuoy E, Langerman J, Clark AT, Jaenisch R, Plath K (2017) Human naïve pluripotent stem cells model X chromosome dampening and X inactivation. Cell Stem Cell 20:87–101
83. Pastor WA, Chen D, Liu W, Kim R, Sahakyan A, Lukianchikov A, Plath K, Jacobsen SE, Clark AT (2016) Naïve human pluripotent stem cells feature a methylation landscape devoid of blastocyst or germline memory. Cell Stem Cell 18:323–329
84. Nakamura T, Okamoto I, Sasaki K, Yabuta Y, Iwatani C, Tsuchiya H, Seita Y, Nakamura S, Yamomoto T, Saitou M (2016) A developmental coordinate of pluripotency among mice, monkeys and humans. Nature 537:57–62
85. Irie N, Weinberger L, Tang WW, Kobayashi T, Viukov S, Manor YS, Dietmann S, Hanna JH, Surani MA (2015) SOX17 is a critical specifier of human primordial germ cell fate. Cell 160:253–268
86. Sasaki K, Nakamura T, Okamoto I, Yabuta Y, Iwatani C, Tsuchiya H, Seita Y, Nakamura S, Shiraki N, Takakuwa T, Yamamoto T, Saitou M (2016) The germ cell fate of cynomolgus monkeys is specified in the nascent amnion. Nature 39:169–185
87. Sasaki K, Yokobayashi S, Nakamura T, Okamoto I, Yabuta Y, Kurimoto K, Ohta H, Moritoki Y, Iwatani C, Tsuchiya H, Nakamura S, Sekiguchi K, Sakuma T, Yamamoto T, Mori T, Woltjen K, Nakagawa M, Yamamoto T, Takahashi K, Yamanaka S, Saitou M (2015) Robust in vitro induction of human germ cell fate from pluripotent stem cells. Cell Stem Cell 17:178–194
88. Kobayashi T, Zhang H, Tang WWC, Irie N, Withey S, Klisch D, Sybirna A, Dietmann S, Contreras DA, Webb R, Allegrucci C, Alberio R, Surani MA (2017) Principles of early human development and germ cell program from conserved model systems. Nature 546:416–420
89. Kobayashi T, Yamaguchi T, Hamanaka S, Kato-Itoh M, Yamazaki Y, Ibata M, Sato H, Lee YS, Usui J, Knisely AS, Hirabayashi M, Nakauchi H (2010) Generation of rat pancreas in mouse by interspecific blastocyst injection of pluripotent stem cells. Cell 142:787–799
90. Cohen MA, Markoulaki S, Jaenisch R (2018) Matched developmental timing of donor cells with the host is crucial for chimera formation. Stem Cell Reports 10:1445–1452
91. Chen Y, Niu Y, Li Y, Ai Z, Kang Y, Shi H, Xiang Z, Yang Z, Tan T, Si W, Li W, Xia X, Zhou Q, Ji W, Li T (2015) Generation of cynomolgus monkey chimeric fetuses using embryonic stem cells. Cell Stem Cell 17:116–124
92. Fang R, Liu K, Zhao Y, Li H, Zhu D, Du Y, Xiang C, Li X, Liu H, Miao Z, Zhang X, Shi Y, Yang W, Xu J, Deng H (2014) Generation of naïve induced pluripotent stem cells from rhesus monkey fibroblasts. Cell Stem Cell 15:488–496
93. De Los Angeles A, Pho N, Redmond DE Jr (2018) Generating human organs via interspecies chimera formation: advances and barriers. Yale J Biol Med 91:333–342
94. De Los Angeles A, Hyun I, Latham S, Elsworth J, Redmond DE Jr (2018) Human-monkey chimeras for modeling human disease: opportunities and challenges. Stem Cells Dev. https://doi.org/10.1089/scd.2018.0162

Chapter 2

Highly Efficient Derivation of Pluripotent Stem Cells from Mouse Preimplantation and Postimplantation Embryos in Serum-Free Conditions

Alejandro De Los Angeles, Daiji Okamura, and Jun Wu

Abstract

Pluripotency refers to the potential of cells to generate all cell types of the embryo proper. Pluripotency spans a spectrum of cellular states. At one polar extreme is naïve pluripotency, which is identified based on the potential to form germline chimeras. At the other polar extreme is primed pluripotency, in which pluripotent cells are primed to differentiate. Mouse naïve PS cells can be derived from preimplantation embryos. Primed epiblast stem (EpiS) cells are typically isolated from epiblasts of early postimplantation mouse embryos. In this chapter, we describe protocols for highly efficient derivation and propagation of murine naïve and primed PS cell lines in serum-free conditions from preimplantation and postimplantation embryos. We describe generation of mouse naïve PS cells using LIF and inhibitors of MEK and GSK3 kinases and of mouse primed PS cells using FGF2 and IWR1 compound which induces the stabilization of Axin proteins.

Key words Pluripotent stem cells, Embryonic stem cells, Epiblast stem cells, Preimplantation epiblast, Postimplantation epiblast, Pluripotency, Region-selective epiblast stem cells

1 Introduction

Pluripotency is an evanescent cellular property during early development that refers to the capacity of epiblast cells to generate all cell types in an adult organism [1]. For more than two decades, scientists have developed various culture conditions that enabled the derivation of pluripotent stem (PS) cell lines from epiblast cells of distinct spatiotemporal origins that maintain the potential to form all cell types indefinitely in the laboratory. Mammalian biology research has been revolutionized by the use of embryo-derived PS cell lines.

In 1981, PS cells were derived from preimplantation murine blastocysts [2, 3]. These stem cells were termed embryonic stem

Insoo Hyun and Alejandro De Los Angeles (eds.), *Chimera Research: Methods and Protocols*, Methods in Molecular Biology, vol. 2005, https://doi.org/10.1007/978-1-4939-9524-0_2, © Springer Science+Business Media, LLC, part of Springer Nature 2019

(ES) cells. The pluripotency of ES cells is compellingly demonstrated by injecting them into preimplantation embryos and observing their extensive contribution to all cell types of the resulting mice, including gametes.

Decades later, scientists first isolated human ES cells from human blastocysts [4, 5]. Human ES cells, like mouse ES cells, maintained a potential to form derivatives of the three germ layers—ectoderm, mesoderm, and endoderm—while also exhibiting unlimited self-renewal capacity.

Despite both having been isolated from the blastocysts, human ES cells exhibited radically distinct properties from mouse ES cells. For example, mouse ES cells have a domed colony morphology, whereas colonies of human ES cells exhibit a flat-disk shape. Moreover, the signaling requirements for maintaining pluripotency in mouse ES cells and human ES cells are different [6–9]. For years, these different properties were ascribed to species-specific differences in the regulation of pluripotency.

In 2007, two independent studies described a distinct PS cell type, termed epiblast stem (EpiS) cells, which could be isolated from mouse postimplantation embryos [10, 11]. EpiS cells exhibit distinct characteristics when compared with paradigmatic mouse ES cells, and interestingly are more akin to human ES cells. For example, EpiS cells also possess a flattened colony morphology and similar dependence on FGF/TGF-beta signaling pathways for self-renewal. EpiS cells are pluripotent but lack the ability to engraft into developmentally earlier blastocysts and form chimeras. The derivation of EpiS cells broadened the concept of pluripotency and led to the realization that pluripotency is not a singular state. The terms "naïve" and "primed" were coined to distinguish mouse ES and EpiS cells and denote relatedness to epiblasts from the preimplantation and postimplantation embryos, respectively [12].

Conventional culture of PS cells has involved with the supplementation of cytokines and growth factors, such as LIF and FGF, serum or serum replacement, and MEF feeders. In 2008, a landmark study showed that it is possible to derive homogeneous cultures of mouse naïve ES cells in a developmental "ground state" by cultivating mouse ES cells in serum-free medium supplemented with inhibitors of MEK and GSK3 kinases. Ground-state culture enabled highly efficient derivation of mouse ES cells from many different genetic backgrounds, including nonpermissive strains [13]. We subsequently demonstrated that primed EpiS cells could also be derived, at 100% efficiency, from both preimplantation and postimplantation embryos of various genetic backgrounds by supplementing serum-free medium with FGF2 and a Wnt inhibitor IWR1 [14]. This study also showed that it was possible to shift the spatial identity of EpiS cells by Wnt inhibition.

The availability of naïve and primed PS cells greatly facilitated the mechanistic studies of molecular events that occur in the early developing embryo. This chapter describes the methods to enable highly efficient derivation of naïve ES cells from preimplantation embryos and primed rsEpiS cells from postimplantation embryos in serum-free conditions.

2 Materials

2.1 Blastocyst Isolation and ES Cell Derivation

1. Timed pregnant female mice (*see* **Note 1**).
2. Pregnant mares' serum gonadotropin (PMSG).
3. Human chorionic gonadotropin (HCG).
4. M2 medium, hyaluronidase.
5. Oil for embryo culture.
6. Acid Tyrode's solution.
7. 0.1% Gelatin solution
8. Ground-state 2iL ES cell culture medium: 490 mL DMEM/F12, 485 mL neurobasal medium, 5 mL N2 supplement, 10 mL B27 supplement, 1× penicillin/streptomycin, 1× NEAA, 1× GlutaMAX, 0.1 μM bME, 10 ng/mL LIF, 1 μM PD0325901, 3 μM CHIR99021.

2.2 Naïve ES Cell Culture

1. 0.05% Trypsin-EDTA or TrypLE.
2. Ground-state 2iL ES cell culture medium.
3. Freezing medium: 50 mL 2iL medium, 40 mL FBS, 10 mL DMSO.

2.3 Epiblast Isolation and EpiS Cell Derivation

1. Timed pregnant female mouse at 5.5 dpc.
2. Dissection medium: DMEM, 10% FBS, 1× penicillin-streptomycin.
3. rsEpiS cell culture medium: 490 mL DMEM/F12, 485 mL neurobasal medium, 5 mL N2 supplement, 10 mL B27 supplement, 1× penicillin/streptomycin, 1× NEAA, 1× GlutaMAX, 0.1 μM bME, 20 ng/mL bFGF, 2.5 uM IWR1.

2.4 Primed EpiS Cell Culture

1. 0.05% Trypsin-EDTA or TrypLE
2. rsEpiS cell culture medium,
3. Freezing medium: 50 mL rsEpiS cell medium, 40 mL FBS, 10 mL DMSO.

3 Methods

3.1 Isolation and Culture of the Preimplantation Blastocysts

3.1.1 Superovulation and Blastocyst Harvest

1. Inject 5 IU PMSG intraperitoneally into female mice.
2. Two days later, inject 5 IU hCG intraperitoneally into female mice.
3. Set up mating of injected female mice with males.
4. One day after injecting mice with hCG, prepare drops of 50 μL M2 and 50 μL M2 plus 0.3 mg/mL hyaluronidase in a 60 mm petri dish. Cover M2 and M2/hyaluronidase drops with oil. Also prepare drops of 50 μL KSOM media, which are also covered with oil. Maintain dishes at 37 °C.
5. Euthanize plug-positive female mice. Pick up the uterus and cut the cervix and between the uterus and the oviduct.
6. Place the uterus in a 60 mm dish. Flush 1 mL of M2 medium through the cervix with syringe and needle, and collect blastocysts with a mouth pipette.
7. Transfer the collected blastocysts into a drop of 20 μL M2 medium covered with oil. Then 2–3 blastocysts are subjected to zona pellucida removal by soaking them in acid Tyrode's solution for 15–30 s followed by sequential wash in three drops of M2 medium. Repeat this step until all the blastocysts are processed.
8. Blastocysts without zona pellucida are placed onto MEFs in 2iL ES culture medium for ES cell derivation, and medium is changed every 2–3 days.

3.2 Naïve ES Cell Culture

3.2.1 Establishment of ES Cell Lines

1. After allowing the blastocyst outgrowth reach a significant size (between 7 and 16 days), prepare for the first passage (Fig. 1).

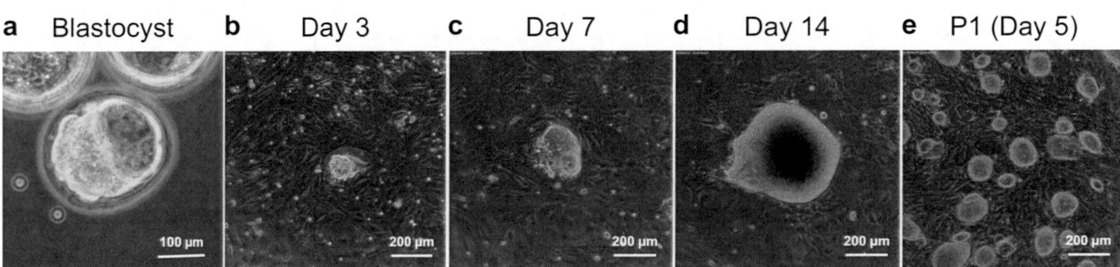

Fig. 1 Naïve ES cell derivation in 2iL. (**a**) Morphology of a blastocyst. (**b**) Morphology of primary preimplantation epiblast colony after 3 days of culture on irradiated MEFs. Notice the domed morphology and sharp defined borders. (**c**, **d**) Morphology of ES cell colony shown in (**b**) after 4 and 11 additional days of culture (7 and 14 days total, respectively). (**e**) Representative image of established naïve ES cells

3.2.2 Passaging of Blastocyst Outgrowth and ES Cells

1. One day before passaging, seed MEFs onto gelatinized plates.
2. Aspirate media from the well containing epiblast outgrowth or ES cells.
3. Add trypsin-EDTA or TrypLE, and incubate in the 37 °C incubator for 5 min.
4. Add serum-containing medium at least at 1:1 ratio (to neutralize trypsin) and finalize dissociation of epiblast outgrowth or ES cells by gentle triturating using a P1000 pipette.
5. Transfer cell suspension to a conical tube and spin cell suspension for 5 min at 200 × *g*.
6. Resuspend cells in an appropriate volume of medium for distribution to new plates seeded with MEFs (1:1 ratio for epiblast outgrowth and 1:20–1:30 for established ES cells).

3.2.3 Cryopreservation of ES Cells

1. Dissociate ES cells as described in Subheading 3.2.2.
2. Collect ES cells by centrifugation at 200 × *g* for 5 min.
3. Resuspend the cell pellet in appropriate volume of freezing medium (e.g., 1 mL for cells collected from one well of a 6-well plate) and aliquot 1 mL of cell suspension into each cryotube.
4. Place cryotubes in cryo-freezing container and store at −80 °C. The following day, transfer vials to a liquid nitrogen freezer for permanent storage.

3.3 Isolation and Culture of Postimplantation Mouse Epiblast

3.3.1 Isolation of Postimplantation Mouse Embryos

1. Euthanize timed pregnant mice (*see* **Note 2**).
2. Transfer uterus to a 100 mm Petri dish containing PBS to wash.
3. Transfer uterus again to a 100 mm Petri dish containing dissection medium.
4. Remove each decidua to reveal egg cylinders using forceps.
5. Transfer each embryo to a 60 mm Petri dish containing dissection medium (Fig. 2).

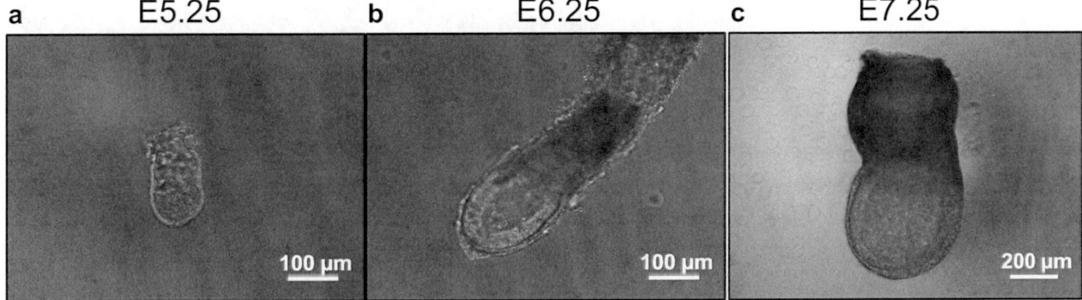

Fig. 2 Postimplantation embryos. (**a**) E5.25 embryo. (**b**) E6.25 embryo. (**c**) E7.25 embryo

3.3.2 Removal of Extraembryonic Tissues

1. Pull back Reichert's membrane.
2. Peel the visceral endoderm away mechanically with a sharpened tungsten needle.
3. Cut embryo at embryonic/extraembryonic boundary and remove extraembryonic fragment containing the extraembryonic ectoderm.
4. The epiblast fragment should remain in the dissection medium.

3.3.3 Postimplantation Epiblast Explant

1. Transfer each epiblast fragment to a single well of a 4-well dish seeded with mitotically inactived MEFs.
2. Cultivate the epiblast fragment in rsEpiS medium (supplemented with FGF2 and IWR1). Medium is changed every other day.

3.4 Primed rsEpiS Cell Culture

3.4.1 Establishment of Region-Selective Epiblast Stem Cell Lines

1. After 4 days of growth, prepare epiblast explant/outgrowth for first passage (Fig. 3).
2. Dissociate the epiblast outgrowths with TrypLE. If clonal rsEpiS cell lines derived from single epiblast cells are desired, *see* **Note 3**.

(1) Visceral endoderm
(2) Ectoplacental cone
(3) Extraembryonic ectoderm
(4) Epiblast

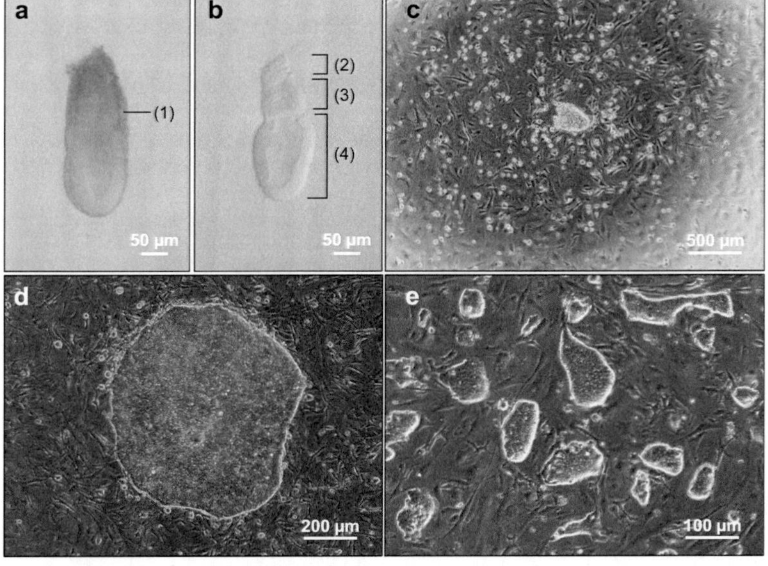

a. E6.5 embryo

b. After removal of visceral endoderm

c. An isolated epiblast on MEFs (Day 0)

d. Day 2 outgrowth of epiblast

e. Established rsEpiSCs at passage 10

Fig. 3 Primed rsEpiS cell derivation (**a**) An E6.5 embryo. (**b**) Embryo after removal of visceral endoderm. (**c**) An isolated epiblast on MEFs (day 0). (**d**) Day-2 outgrowth of epiblast. (**e**) Established rsEpiS cells at passage 10

3. Transfer epiblast clusters to a well freshly seeded with irradiated MEFs. Cultivate in rsEpiS cell medium. Medium is changed daily.

3.4.2 Passaging of rsEpiS Cells

1. In our laboratory, rsEpiS cells are generally cultured in 6-well plates and passaged every 3–4 days. rsEpiS cells can be passaged as single cells enzymatically. For passaging, aspirate rsEpiS cell medium, add 0.5 mL of TrypLE, and incubate at 37 °C for 5–10 min.
2. Add 0.5 mL of rsEpiS medium to each well and dissociate rsEpiS cell colonies by gentle pipetting up and down 8–10 times. Transfer cell suspensions into a 15 mL conical tube.
3. Centrifuge cell suspension at 200 × *g* for 5 min.
4. Aspirate the supernatant. Resuspend the cell pellet in an appropriate volume that can be redistributed to additional tissue culture plates (passage ratio 1:50).
5. EpiS cell medium is changed on a daily basis.

3.4.3 Cryopreservation of rsEpiS Cells

1. Dissociate rsEpiS cells as described in Subheading 3.4.2.
2. Transfer to a conical tube and spin down at 200 × *g* for 5 min.
3. Resuspend the cell pellet in appropriate volume of freezing medium (e.g., 1 mL for cells collected from one well of a 6-well plate) and aliquot 1 mL of cell suspension into each cryotube.
4. Place cryotubes in cryo-freezing container and store at −80 °C. The following day, transfer vials to a liquid nitrogen freezer for permanent storage.

4 Notes

1. Timed pregnant mice can be ordered from commercial suppliers or timed mating can be set up in the laboratory. Embryonic day 0.5 (E0.5) of pregnancy refers to the noon of the presence of a vaginal plug after overnight mating.
2. For rsEpiS cell derivation, embryo collection can be done at developmental stages between E5.25 and E7.75. In addition, rsEpiS cell lines can be directly derived from isolated ICMs from E3.5 blastocysts.
3. For derivation of clonal rsEpiS cell lines, day-4 outgrowths of isolated epiblasts or established rsEpiS cell lines are treated with trypsin-EDTA for 10 min at 37 °C, followed by repeated pipetting (~40 times) with a P200 pipette. Dissociated cells are passed through a 40 μM cell strainer to obtain a single-cell suspension and cultured on MEFs at clonal density (e.g., 500 cells in one well of a 6-well plate) in rsEpiS cell medium. Single colonies can be manually picked for further cultivation.

References

1. De Los Angeles A, Ferrari F, Xi R, Fujiwara Y, Benvenisty N, Deng H, Hochedlinger K, Jaenisch R, Lee S, Leitch HG, Lensch MW, Lujan E, Pei D, Rossant J, Wernig M, Park PJ, Daley GQ (2015) Hallmarks of pluripotency. Nature 525(7570):469–478
2. Evans MJ, Kaufman MH (1981) Establishment in culture of pluripotential cells from mouse embryos. Nature 292:154–156
3. Martin GR (1981) Isolation of a pluripotent cell line from early mouse embryos cultured in medium conditioned by teratocarcinoma stem cells. PNAS 78:7634–7638
4. Thomson JA, Itskovitz-Eldor J, Shapiro SS, Waknitz MA, Swiergiel JJ, Marshall VS, Jones JM (1998) Embryonic stem cells derived from human blastocysts. Science 282:1145–1147
5. Reubinoff BE, Pera MF, Fong CY, Trounson A, Bongso A (2000) Embryonic stem cell lines from human blastocysts: somatic differentiation in vitro. Nat Biotechnol 18:399–404
6. Ying QL, Nichols J, Chambers I, Smith A (2003) BMP induction of Id proteins suppresses differentiation and sustains embryonic stem cell self-renewal in collaboration with STAT3. Cell 115:281–292
7. Vallier L, Alexander M, Pedersen RA (2005) Activin/Nodal and FGF pathways cooperate to maintain pluripotency of human embryonic stem cells. J Cell Sci 118:4495–4509
8. Ying QL, Wray J, Nichols J, Batlle-Morera L, Doble B, Woodgett J, Cohen P, Smith A (2008) The ground state of embryonic stem cell self-renewal. Nature 453:519–523
9. Wu J, Belmonte JCI (2016) Stem cells: a renaissance in human biology research. Cell 165:1572–1585
10. Tesar PJ, Chenoweth JG, Brook FA, Davies TJ, Evans EP, Mack DL, Gardner RL, McKay RD (2007) New cell lines from mouse epiblast share defining features with human embryonic stem cells. Nature 448:196–199
11. Brons IG, Smithers LE, Trotter MW, Rugg-Gunn P, Sun B, Chuva de Sousa Lopes SM, Howlett SK, Clarkson A, Ahrlund-Richter L, Pedersen RA, Vallier L (2007) Derivation of pluripotent epiblast stem cells from mammalian embryos. Nature 448:191–195
12. Nichols J, Smith A (2009) Naïve and primed pluripotent states. Cell Stem Cell 4:487–492
13. Wu J, Izpisua Belmonte JC (2015) Dynamic pluripotent stem cell states and their applications. Cell Stem Cell 17:509–525
14. Wu J, Okamura D, Li M, Suzuki K, Luo C, Ma L, He Y, Li Z, Benner C, Tamura I, Krause MN, Nery JR, Du T, Zhang Z, Hishida T, Takahashi Y, Aizawa E, Kim NY, Lajara J, Guillen P, Campistol JM, Esteban CR, Ross PJ, Saghatelian A, Ren B, Ecker JR, Izpisua Belmonte JC (2015) An alternative pluripotent state confers interspecies chimeric competency. Nature 521:316–321

Chapter 3

Derivation of Hematopoietic Stem and Progenitor Cells from Human Pluripotent Stem Cells

Ryohichi Sugimura

Abstract

Hematopoietic stem cell (HSC) transplantation is a curative treatment for hematologic malignancies and innate immunodeficiency, but its applications are limited to matched donors, and the availability of umbilical cord blood of immune types. Therefore, derivation of HSCs from patient-specific human induced pluripotent stem cells (hiPSCs) is a holy grail in regenerative medicine. However, derivation of HSCs from iPSCs has proven elusive. Here, the authors developed a method to derive hematopoietic stem and progenitor cells (HSPCs) by combining two established methods. The first method mimics embryonic development by directed differentiation of iPSCs into a cellular state termed hemogenic endothelium (HE) by stepwise exposure to different combinations of morphogens and cytokines. In the second method, transcription factors are induced in HE cells for specification into HSCs. By combining these approaches, the authors identified a set of select transcription factors that programmed HE cells into HSPCs with long-term and multilineage capacity in vivo. In this chapter, I provide an overview of this technology, technical tips, and future applications.

Key words Pluripotent stem cells, Gene regulatory network, Hematopoietic stem cells

1 Introduction

Decades of research attempted to induce HSCs from human pluripotent stem cells by mimicking embryonic development through stepwise exposure to morphogens [1–3]. HSCs arise from the ventral wall of the dorsal aorta of the embryo. The specialized vascular endothelial cells (hemogenic endothelial cells) convert to HSCs, and then migrate to and settle down in bone marrow via circulation. The identity and regulation of hemogenic endothelial cells are well described in mouse embryos, but not as much precisely studied in human. Signals from vascular niche involved in the development of human HSCs are still not clear. To induce HSCs from human pluripotent stem cells, the important step is to clarify the signals involved in specification and conversion of hemogenic endothelial cells to HSCs in human embryos.

Insoo Hyun and Alejandro De Los Angeles (eds.), *Chimera Research: Methods and Protocols*, Methods in Molecular Biology, vol. 2005, https://doi.org/10.1007/978-1-4939-9524-0_3, © Springer Science+Business Media, LLC, part of Springer Nature 2019

In this chapter, we outline the principle, current limitation, and potential applications of the protocol. Detailed methods are described in an accompanying protocol (Nature Protocol Exchange, http://www.nature.com/protocolexchange/protocols/5913). The first step of the protocol begins with embryoid body (EB) formation from human iPSCs and generation of hemogenic endothelial (HE) cells [4, 5]. HE cells are treated with small molecules and morphogens to induce endothelial hematopoietic transition (EHT), as well as transduced with lentiviral vectors encoding transcription factors. Transduced HE cells are transplanted into sublethally irradiated immunodeficient mice. Transplanted mice are treated with doxycycline to induce expression of transgenes for 2 weeks. 6–12 weeks after injection, engraftment of transplanted cells is analyzed from either peripheral blood or bone marrow. The timeframe of each experiment will be total 12–14 weeks: EB formation (8 days), EHT (3 days), transcription factor transduction (1 day), transgene induction (2 weeks), and engraftment (−10 weeks).

1.1 Preparation of Human PSCs

Human iPSCs are maintained on mouse embryonic fibroblast (MEF) feeders. It is crucial to maintain a high quality of stem cell lines by monitoring the morphology of colonies. Approximately 20 dishes of 10 cm plates are needed for each transplant experiment. 100,000–300,000 transduced HE cells are transplanted into each mouse and roughly 1–3 mice are needed per experiment.

1.2 Embryoid Body (EB) Formation

Our experimental system mimics embryonic development by stepwise exposure to different combinations of morphogens and cytokines. HE cells are isolated by magnetic activated cell sorting (MACS) at day 8 of EB formation. It is recommended to analyze the proportion of HE cells (CD34-positive FLK1-positive cells) post-MACS isolation. It is advised to use batches of over 80% enrichment of HE cells for transplantation.

1.3 Endothelial to Hematopoietic Transition (EHT)

CD45+ hematopoietic cells appear from CD34+ HE cells during exposure with hematopoietic cytokines. The presence of EHT is the most important factor for predicting whether the transplantation experiment will succeed. In our experience, engraftment can be only achieved in cases where EHT occurred within 3 days from the isolation of HE cells. The key experimental parameters for induction of EHT are (1) quality of iPSCs and (2) purity of HE cells after isolation (80% or more is recommended).

1.4 Induction of Transcription Factors

Transcription factors are introduced by a two-vector tet-ON polycistronic lentiviral system. One vector expresses LCOR-HOXA9-HOXA5 and the other vector expresses RUNX1-ERG. The desired multiplicity of infection (MOI) for each vector is 2.0.

1.5 Transplantation

Cells are harvested and transplanted 24 h after viral infection. The authors do not re-isolate CD34-positive cells at this stage. We injected 100,000–300,000 cells into each irradiated NSG immunodeficient mice intrafemorally. Doxycycline is administered to mice for 2 weeks to induce transcription factors.

1.6 Engraftment Assay

Bone marrow, spleens, and thymuses from recipient mice are collected and analyzed using flow cytometry. Since human blood cells are difficult to peripherize, we recommend analyzing the above-mentioned organs at 8–12 weeks after transplantation and using cord blood-injected mice as a control.

2 Results

After 8–12 weeks of transplantation, human blood cells can be found in the bone marrow of the transplanted mouse. Approximately 0.1–20% of human chimerism in bone marrow with myeloids, erythrocytes, B cells, and few T cells can be observed. In order to prove that hematopoietic stem/progenitor cells were generated, the following two experiments should be conducted: (1) confirmation of self-renewal ability of hematopoietic stem/progenitor cells by secondary transplantation and (2) analysis of genomic integration sites of the lentivirus confirming the clonality of HSCs. Furthermore, to assess the functionality of engrafted human immune cells, the following three experiments would be conducted. (1) T-cell response ability by interferon γ production upon stimulation; (2) acquisition of immunological diversity of T cells by sequence analysis of T-cell receptor; and (3) B-cell responsiveness by immunization. In our experience, hematopoietic stem/progenitor cells can be induced with similar efficiency from hiPSCs and hESCs.

3 Comparison Between Other Methods

1. Generation of HSPCs using teratoma formation: As the first demonstration of deriving hematopoietic stem/progenitor cells from human pluripotent stem cells, CD34-positive cells from teratoma were engrafted with multilineage capacity in murine recipients [6, 7]. However, there are two caveats in this method: (1) generation of CD34+ cells were rare in teratoma and (2) there is potential safety issue in transplanting tumor-derived cells.
2. Direct reprogramming of vascular endothelial cells to HSPC-like cells: Rafii's group has shown that hematopoietic stem/progenitor cell can be derived both from mouse and human

endothelial cells by introducing four transcription factors [8, 9]. However, co-culture and co-transplantation with immortalized human vascular endothelial cell line hampers convenience and safety of future application.

4 Current Limitations and Venues for Improvement

While these experiments provide proof-of-principle evidence that one could derive HSPCs from human PSCs, this technology is still at its infancy. To achieve full realization of this technology, further optimizations are necessary. Two issues to be resolved are as follows: (1) Potential risks associated with viral vectors: Integration-associated insertional mutagenesis may predispose toward cancer. Although we have not observed the development of malignancy in any transplanted mice, the establishment of an integration-free vector system, such as episomes or self-replicating RNAs, will reduce or eliminate the risks associated with insertional mutagenesis. (2) Increasing efficiency: As described above, engraftment events are still rare and highly depend on whether EHT occurs. (3) Optimization of transcription factor induction regimen: Currently, all transcription factors are expressed at the same time and turned off 2 weeks after transplantation. But each transcription factor may work differently. Temporal manipulation of transgene expression might improve the efficiency of HSC derivation. (4) Identification of molecular mechanisms: How transcription factors drive the acquisition of hematopoietic stem/progenitor identity is of fundamental interest. Taken together, there exists a plethora of future opportunities to modify and improve the generation of HSPCs from human PSCs.

5 Future Directions and Applications

1. Disease modeling such as congenital anemia or immunodeficiency: In vitro drug screening will be the most practical application.
2. Off-the-shelf blood cells: Platelets and erythrocytes, that are anuclear and free from transformation, hold promises for cell therapy. Efficient production of platelets and erythrocytes would be achieved by hPSC-derived HSCs.
3. Establishment of integration-free system: It is highly demanded both for safe supply of cells and genuine demonstration of full reprogramming to HSPCs independent from remaining transgenes. Either episomes or self-replicating RNA system, with the latter totally eliminating risks of genomic integration, will be developed.

4. Establishment of hematopoietic stem/progenitor cells in vitro: The current method depends on transgene induction in vivo, presumably depending on microenvironmental cues to specify HSCs. We need to identify and convey new signaling factors that enable derivation of HSCs fully in vitro.

References

1. Kennedy M et al (2012) T lymphocyte potential marks the emergence of definitive hematopoietic progenitors in human pluripotent stem cell differentiation cultures. Cell Rep 2:1722–1735. https://doi.org/10.1016/j.celrep.2012.11.003
2. Pereira CF et al (2013) Induction of a hemogenic program in mouse fibroblasts. Cell Stem Cell 13:205–218. https://doi.org/10.1016/j.stem.2013.05.024
3. Riddell J et al (2014) Reprogramming committed murine blood cells to induced hematopoietic stem cells with defined factors. Cell 157:549–564. https://doi.org/10.1016/j.cell.2014.04.006
4. Ditadi A et al (2015) Human definitive haemogenic endothelium and arterial vascular endothelium represent distinct lineages. Nat Cell Biol 17:580–591. https://doi.org/10.1038/ncb3161
5. Ditadi A, Sturgeon CM (2016) Directed differentiation of definitive hemogenic endothelium and hematopoietic progenitors from human pluripotent stem cells. Methods 101:65–72. https://doi.org/10.1016/j.ymeth.2015.10.001
6. Suzuki N et al (2013) Generation of engraftable hematopoietic stem cells from induced pluripotent stem cells by way of teratoma formation. Mol Ther 21:1424–1431. https://doi.org/10.1038/mt.2013.71
7. Amabile G et al (2013) In vivo generation of transplantable human hematopoietic cells from induced pluripotent stem cells. Blood 121:1255–1264. https://doi.org/10.1182/blood-2012-06-434407
8. Sandler VM et al (2014) Reprogramming human endothelial cells to haematopoietic cells requires vascular induction. Nature 511:312–318. https://doi.org/10.1038/nature13547
9. Lis R et al (2017) Conversion of adult endothelium to immunocompetent haematopoietic stem cells. Nature 545:439–445. https://doi.org/10.1038/nature22326

Chapter 4

Cancer Stem Cells: Concepts, Challenges, and Opportunities for Cancer Therapy

May Yin Lee, Rajshekhar R. Giraddi, and Wai Leong Tam

Abstract

Cancer stem cells (CSCs) are a subpopulation of cancer cells with self-renewal capacity, that fuel tumor growth and contribute to the heterogeneous nature of tumors. First identified in hematological malignancies, CSC populations have to date been proposed in solid tumors in various organs. In vitro and in vivo assays, mouse genetic models, and more recently single-cell sequencing technologies and other '-omics' methodologies have not only facilitated the identification of novel CSC populations but also revealed and clarified novel properties of CSCs. Increasingly, both cell-autonomous and CSC niche factors are recognized as important contributors of CSC properties. The deepened understanding of CSC properties and characteristics would enable and facilitate the rational design of CSC-specific therapeutics that would, ideally, have high selectivity for cancer cells, eliminate tumor bulk, and prevent tumor recurrence. Addressing these issues would form some of the key challenges of the CSC research field in the coming years.

Key words Cancer stem cells, Tumorigenesis, Cellular plasticity, Tumor heterogeneity, Drug resistance, Targeted therapy

1 Introduction

Over the past decades, a subpopulation of cells within cancers termed cancer stem cells (CSCs) have become an obsession for many researchers. In this chapter we give a background to this subject of intense research drawing from historical perspectives, models, key features, and techniques of study of CSCs and finally offer some future research directions for the field that would be key to understand CSCs and their implications for cancer therapy.

2 Historical Background, Features, and Controversies of CSCs

Tumors are made up of a myriad of cells that have functional and phenotypic heterogeneity. One of the factors to account for the functional and phenotypic diversity is a distinct population of self-

Insoo Hyun and Alejandro De Los Angeles (eds.), *Chimera Research: Methods and Protocols*, Methods in Molecular Biology, vol. 2005, https://doi.org/10.1007/978-1-4939-9524-0_4,

renewing malignant cell population termed cancer stem cells (CSCs). CSCs are distinguished by the ability to self-renew and having the developmental potential to recapitulate a variety of cell types found in a tumor.

The first ideas of CSCs came from the hematopoietic system. Upon the acquisition of genetic mutations, normal hematopoietic stem cells can give rise to leukemic stem cells which are cancer-initiating cells of acute myeloid leukemia (AML) [1, 2]. From these studies, a cellular hierarchy in tumorigenesis was first proposed where CSCs at the apex can give rise to more differentiated and short-lived progeny. Of note, while there is clear evidence of the presence of CSCs in tumors arising in blood cell lineages that mirror normal developmental cues, there is considerable debate on the presence of CSCs in solid tumors. This is, in part, due to the technical challenges to isolate and characterize functional CSCs (discussed below) in these tumors. Nonetheless, at present, many studies have identified putative CSCs in a variety of solid tumors such as the breast, colon, brain, skin, and intestine (Table 1). Notably, even within a single tumor, several distinct CSC populations may exist [3, 4].

Two main models (and their variations) (Fig. 1) have been proposed to describe tumor growth and acquisition of heterogeneity. The hierarchical model, mediated by single or multiple CSCs, assumes a fixed or rigid cellular hierarchy akin to normal development. At the top of the hierarchy is the CSC. Below that, at each level of the hierarchy, cells can gain or lose mutation(s) that will result in the formation of differentiated and heterogeneous clones. A stable CSC phenotype suggested in this model precludes the stochastic interconversion of stem-like cells and more differentiated cell types.

A more fluid or dynamic model termed the stochastic or clonal evolution model suggests that cells can enter and exit the stem cell state depending on intrinsic factors like genetic evolution, cell-state changes, and extrinsic microenvironmental stimuli. This model suggests that all cells would have the same tumor-initiating capacity, yet tumor-initiating capacity is stochastically restricted to a subset of cells within the tumor population. With appropriate stimuli (e.g. gain of an advantageous mutation), any cell can potentially give rise to a dominant clone within the tumor.

It is pertinent to note that these models are not mutually exclusive and can be viewed as integrated processes. Stochastic events could allow cells to dedifferentiate to become CSCs and which would then give rise to hierarchically organized cell populations. These processes could be reiterated during the course of tumor evolution.

Several features of CSCs deserve clarification here. To the uninitiated, the term CSC may evoke some assumptions that are not necessarily warranted. Indeed, the term suggests that the cell of

Table 1
CSC markers in specific tumor types

Tumor type	CSC marker	References
Acute myeloid leukemia	$CD34^+$ $CD38^-$, $CD47^+$, CCL-1^+, $CD96^+$, $TIM3^+$, $CD32^+$, $CD25^+$, $ALDH^+$	[1, 2, 80, 98–101]
Non-Hodgkin lymphoma	$CD47^+$	[81]
Bladder	EMA^- $CD44v6^+$	[102]
Bone sarcoma	STRO-1^+ $CD105^+$ $CD44^+$	[103]
Breast	$CD44^+$ $CD24^{-/low}$, $EpCAM^+$, $ALDH^+$	[104]
Brain	$CD133^+$	[105]
Colorectal	$CD133^+$, $EpCAM^+$, $Lgr5^+$, $CD166^+$, $CD44^+$,$CD24^+$	[5, 106–108]
Gallbladder	$CD44^+$ $CD133^+$	[109]
Gastric	$CD44^+$ $CD24^+$	[110, 111]
Head and neck	$CD44^+$, $CD24^+$	[112, 113]
Liver	$CD133^+$, $CD44^+$ $CD90^+$	[114–116]
Lung	Sca-1^+ $CD45^-$ PECAM-1^- $CD34^+$, $CD133^+$	[117]
Melanoma	$CD20^+$	[118]
Ovarian	$CD44^+$ $CD117^+$	[119–121]
Pancreatic	$CD24^+$ $CD44^+$ $EpCAM^+$	[122]
Prostate	$CD44^+$ $\alpha_2\beta_1{}^+$ $CD133^+$, $EpCAM^+$, Sca-1^+	[123]
Renal	$CD105^+$	[124, 125]
Skin SCC	$CD34^+$, $Sox2^+$	[8, 23, 126]
Skin BCC	$Sox9^+$	[127]
Intestinal	$Lgr5^+$	[5]
Esophageal	$CD90^+$	[128]

origin of CSCs is *necessarily* a normal stem cell as in the case of AML [1, 2], intestinal cancer [5], and basal cell carcinoma [6]. This notion, however, is being steadily disproved in various liquid and solid tumor models which show that non-stem cells or differentiated progenitor cells may adopt a CSC phenotype and contribute to tumor growth and bulk as suggested by the stochastic model [7–9]. These observations highlight the plasticity of cellular phenotypes during tumorigenesis

In relation to normal stem cell characteristics, CSCs have been thought to be quiescent and divide asymmetrically. The

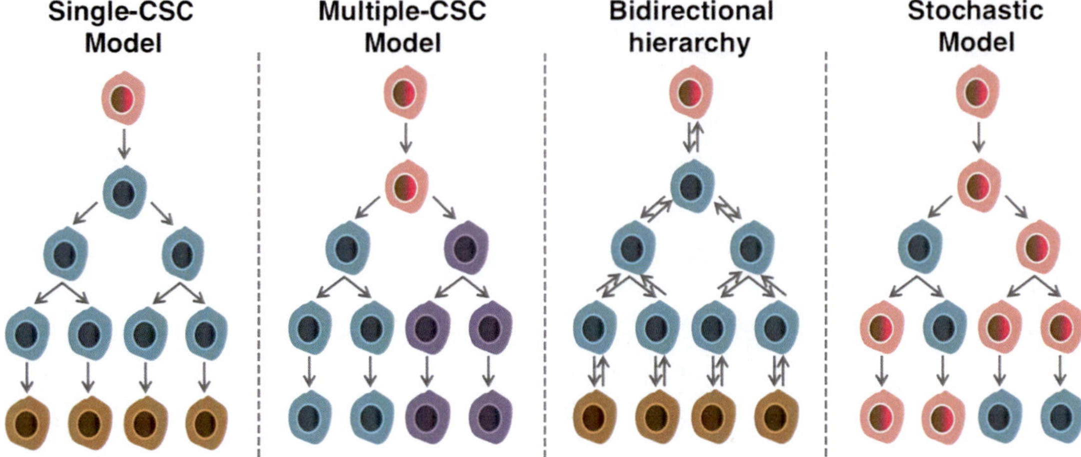

Fig. 1 CSC models: Tumors can be propagated by a single or multiple CSCs (pink cell) that produces single or multiple progenitors/transint amplifying cells (blue and purple cells) and differentiated cells (brown cells) in a unidirectional or bidrectional hierarchy. In some tumors, all cells display plasticity and are able to produce all cell types in the tumor (stochastic model). This implies that the overall CSC content may change at all times. The generation of various CSC progenies is determined by intrinsic and extrinsic factors

experimental evidence supporting these traits is few and warrants further clarification due to their implications on therapy (*see* section below).

Another frequently debated issue is the frequency of CSC in a given tumor which is usually assumed to be small akin to the frequency of stem cells in normal tissues. While many studies show a low frequency of CSCs in tumors, at least a few observations counter this claim. When lymphomas and leukemias of mouse origin are transplanted into histocompatible mice, a very high frequency (at least one in ten) of the tumor cells can seed tumor growth [10]. In melanoma, it has been proposed that CSC frequency can vary between 2.5 and 41% [11].

3 Methods to Study CSCs

The CSC field has traditionally relied on a few methods to study CSCs, oftentimes drawing inspiration from the hematopoietic stem cell field which has pioneered most of them (Table 2). The purification and isolation of CSCs by specific cell surface marker expression in AML have shown that a particular subset of cells in the hematopoietic niche characterized by $CD34^{+}$ $CD38^{-}$ was able to give rise to AML [1, 2]. Attempts to isolate CSCs from solid tumors such as the breast, colon, brain, intestine, and skin by cell surface marker expression or immunophenotyping have proven to be more challenging. Part of the challenge and difficulty is the

Table 2
Methods to identify, isolate, and assess CSCs

Techniques	CSC characteristic assessed	Advantages	Disadvantages	Model systems
Immunophenotyping/ cell surface marker expression	Stemness	Easy, fast Efficient way to isolate putative CSCs	Limited applicability, high variability	Cell lines, tumors
Hoechst dye exclusion, side population	Quiescence	Easy, fast	Limited applicability	Cell lines, tumors
ALDH activity	Drug resistance	Easy, fast	Limited applicability, high variability	Cell lines, tumors
Limiting dilution transplantation and serial passaging	Tumor initiation	Ability to assess tumor heterogeneity Quantitative measure/ estimate of CSC frequency	Expensive Reliance on immunocompromised models	Cell lines, tumors
Sphere formation	Self-renewal	Easy Ability to derive patient models	Limited applicability	Cell lines, tumors
Lineage tracing	Self-renewal, clonogenicity	In vivo context Ability to study long-term repopulation	Expensive, heavy reliance on organism-specific genetic models	Model organisms (e.g. mouse, zebrafish)
Single-cell profiling	Cellular hierarchy	Unbiased Ability to assess tumor heterogeneity	Expensive Heavy reliance on computational methods and power	Cell lines, tumors

observation that there is no one universal CSC marker for all tumors from various organs. Moreover, CSC marker expression could change with tumor stage and evolution due to cellular and phenotypic plasticity [12].

However, several more common markers may be helpful in initial assessments of CSC activity such as CD44, CD133, and EpCAM (*see* Table 1 for a list of CSC markers). Another intriguing marker usually used to assess CSC activity is ALDH which was originally identified in hematopoietic stem cells [13]. ALDH activity is linked to drug resistance. Several studies have pointed to ALDH high activity as a marker for CSCs in malignant human breast stem cells [14] and many other solid tumors. Like other molecular markers, ALDH expression may change during the course of tumor evolution [15].

Besides the immunophenotyping techniques described above, the self-renewal capacity of CSCs is typically assessed by the ability of cells to form spheres in vitro [16]. Many efforts have been devoted to the derivation of 3D tumor sphere models from a variety of organs to study CSCs. Our own work shows that in non-small cell lung carcinoma (NSCLC), tumor spheres have enhanced CSC or tumor-initiating characteristics compared to their isogenic adherent cell line counterparts (Wang et al., manuscript in review).

CSCs are defined by their clonal long-term repopulation capacity. As such, candidate CSCs can also be assessed by limiting-dilution transplantations, oftentimes with supportive materials such as Matrigel, in immunocompromised mouse hosts. The assay tests for the ability of transplanted cells, at low or clonal densities, to form a tumor in vivo and recapitulate the phenotypic and heterogeneity of the parental tumor. By definition, CSCs should be able to self-renew and initiate tumors in multiple rounds of in vivo serial passaging. In the absence of other corroborating evidence drawn from other assays, this particular feature has led some in the field to adhere to the more puristic terms 'tumor-initiating cells' (TICs) or 'tumor-propagating cells' over the more contentious CSCs. Moreover, it has been proposed that the term CSC should be restricted to cells that can be prospectively isolated from tumors. If this is not the case, the functional term TIC is perhaps more appropriate.

The limiting-dilution transplantation method, when performed with a range of cell doses, can also be used to estimate the frequency of CSCs and compare the 'stemness' of various CSC populations within or across tumors. However, this approach suffers from the fact that the transplantation site, typically subcutaneous, may not fully recapitulate the native tumor microenvironment essential for tumor growth and thereby produce artificial selection biases. For example, while primary serous ovarian cancer contains a large proportion of $CD133^+$ CSCs, a significant number of xenografted tumors in the mammary fat pad contain a large proportion of $CD133^-$ TICs [17]. It is therefore imperative that researchers test a variety of transplantation sites and methods when assaying CSCs.

Techniques that rely on isolating and studying CSCs in vitro, though informative, may mask or lead to biases in CSC characteristics observed in vivo. Studies of normal mammary stem cells have shown that while a single basal cell can give rise to the entire epithelial network of the mammary gland ex vivo [18, 19], basal cells are shown to be unipotent in vivo using lineage tracing methods [20, 21]. Van Keymeulen and Lee et al. [22] show by in vivo lineage tracing that basal and luminal cells only become multipotent upon the overexpression of a potent oncogene like PIK3-CA^{H1047R} or in combination with p53 loss-of-function. This and other observations highlight the importance of niche and microenvironment-specific signals that govern cellular fates (*see* the section below on intrinsic and extrinsic factors).

The caveats of studying CSCs in vitro have intensified efforts in developing new animal models and in vivo lineage tracing methods which are touted as the gold standard for studying CSCs. In lineage tracing, a cell or populations of cells are marked by a promoter-specific transgenic reporter. Upon genetic recombination, the transgenic reporter, typically a fluorescent marker or lacZ, is transmitted to all of the parental cell's progeny. This would allow the establishment of a cellular hierarchy in vivo. Using this method, Boumahdi et al. [23] show that basal skin cells expressing Sox2 represent a CSC population in squamous cell carcinoma. Importantly, tumors were eradicated upon lineage ablation of the $Sox2^+$ population, demonstrating that $Sox2^+$ cells make up an important cell lineage that sustains tumor growth. Similarly, using lineage retracing methods, the Clevers group showed the important contribution of $Lgr5^+$ cells in intestinal and colorectal tumors in vivo [5]. Likewise, $Nestin^+$ cells make up a substantial part of glioblastoma [24].

In the absence of markers or when markers are unstable, unbiased lineage tracing methods can be particularly informative to track CSCs and to observe clonal cell growth over a long time frame. Zomer et al. [25] performed unbiased lineage tracing in a mouse model of breast cancer to characterize the nature of the tumor growth and identified the presence of CSCs. Using a Rosa26-Cre Confetti reporter that effectively labels any cell in the tumor independently of biased cell marker expression, researchers detected the presence of large unicolor cell progenies in vivo, suggesting the contribution of single-cell populations. Intravital imaging further confirmed the presence of these large clones in the same mice that had multicolor smaller clones at the start of the recombination. These results suggest that while multiple cell types may initially contribute to tumor propagation at varying levels, a single population ultimately dominates.

Using CA-30 somatic based mutation tracing, the presence of CSC-mediated tumor propagation was also observed in vivo in mouse intestinal adenomas [26]. Depending on the number of CSCs present and the size of the adenomas analyzed, small or large clones emerging from single cells were observed and remained stable for more than one year suggesting the long-term contribution of CSC that fuels the growth of the adenomas in vivo. Such somatic based lineage tracing techniques could be useful to study CSCs in human tumors. Indeed, a combination of nuclear and mitochondrial DNA lesions and methylation patterns elucidated the clonal dynamics of human colorectal adenomas [27]. In addition, CRISPR/Cas9-based lineage tracing methods that are widely applicable in various model systems could be powerful tools for future research.

4 Single-Cell Analysis of CSCs

To date, our understanding of CSCs has been mostly derived from studying cells in bulk, obscuring observations that may be gathered from analyses at the single-cell level. With the advances in single-cell analyses, transcriptomic, epigenetic, and protein expression profile differences between two cells are being captured at the highest resolution and we are now in a better position to clarify the concepts of 'cell states', 'cell fate', and 'cell potential' (Fig. 2). In order to define the presence of distinct CSCs and differentiate it from non-CSCs, it is critical to understand whether CSCs and non-CSCs that may present themselves as distinct cell types (based on the expression of a surface markers) have (1) dynamic and oscillating genetic programs suggesting a high chance of interconversions and plasticity (cell states), (2) are separated by refined and stable genetic programs suggesting a nonconvertible and compartmentalized cell types (cell fate), or (3) respond to activating or inactivating stimuli enabling them to contextually produce cells of all lineages in the tumor (cell potential). While current bulk cell analyses can display differences between CSCs and non-CSCs, minor oscillating gene expression levels, chromatin configurations, and additional regulatory measures during mRNA translation steps must be characterized at the single-cell level to determine if the tumor is governed by stable or unstable CSC population(s). This would allow the identification of irreversible or permanent phenotypic markers for the prospective isolation and characterization of CSCs. Prior to single-cell assays, a landmark study published in 2008 that analyzed differential levels of Sca-1 expression in cells by flow cytometry suggested that what was previously described as

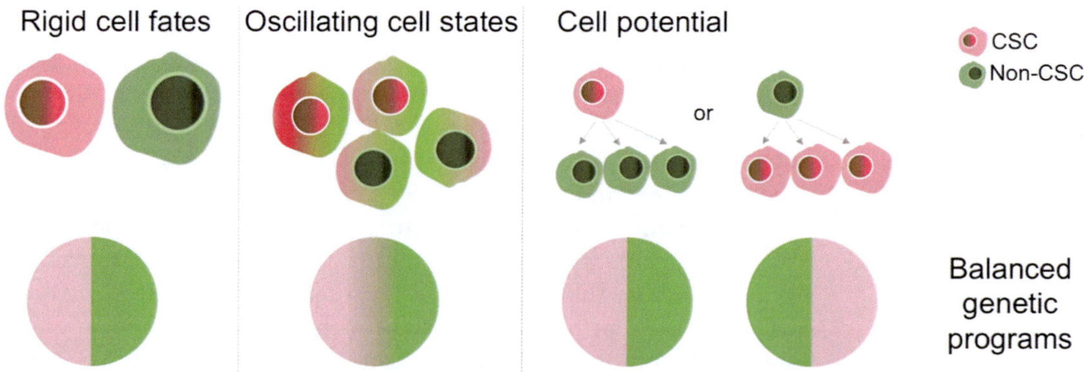

Fig. 2 Identification of CSCs or non-CSCs: Differences between two cells could occur as a result of differential cell fates (with green and red cells displaying distinct genetic programs), cell states (where most cells share genetic programs common to green and red cells but a balanced expression favors cells biased toward either green or red identities), or cell potential (where the distinct red or green cells can generate other lineages of each other depending on the external stimuli or the assay used to test the potentials)

distinct progenitor populations based on Sca-1 low or Sca-1 high expressions were not two distinct cell types but were the same cells at different stages of the cell cycle that exhibited an unsynchronized transcriptional machineries leading to varying levels of Sca-1 [28].

In human glioblastomas, single-cell analyses revealed new transcriptional programs that were different from the dominant transcriptional program obtained using bulk analyses and proposed the presence of the hybrid cell states and their impact on patient prognosis [29]. In addition, single-cell profiling of colorectal tumors identified the presence of new FAP^{-ve} stromal fibroblast that may not respond to potential FAP-directed therapies but aid in paracrine TGF-β signaling for tumor cell survival and propagation. Rather provocatively, this report showed that EMT signatures were enriched in fibroblasts compared to epithelial tumor cells [30], an observation that would have been masked by bulk tumor analyses.

Put together, these findings and many other studies in normal homeostasis and cancer (reviewed in [31]) continue to stir debates on how to view phenotypic and functional cell types in normal and neoplastic samples and the subsequent definitions for cell states, cell fate, and cell potential. Thus, in addition to bulk cell analyses, single-cell assays are and will be enormously useful to understand the biology of tumor heterogeneity, identify and characterize novel cell types, and identify potential therapeutic targets.

5 EMT and CSCs

EMT and CSCs have been inextricably linked in many instances. During cancer evolution, cells may, from a polarized epithelial organization, adopt an undifferentiated, migratory, and invasive mesenchymal cell state. This process mirrors the morphogenetic events that occur during embryonic development. Intriguingly, tumor cells have been shown to exhibit features of embryonic cells [32, 33]. In cancer, EMT occurs upon overexpression of classical transcription factors Zeb1, Twist1, Snail, and Slug, leading to the gain of stem-cell-like properties, tumorigenicity [34–36], and formation of metastasis [37]. Other co-activators may be involved in this process. For example, together with Slug, Sox9 has been shown to be an important co-activator to enhance EMT and promote metastasis in breast cancer cell line xenografts [38]. In addition, the activation of signaling pathways (e.g. TGF-β, FGF, EGF, HGF, Wnt/β-catenin, and Notch) and hypoxia may induce EMT.

The potential to undergo EMT may differ between cell types, leading to differential CSC features. While skin squamous cell carcinomas (SCC) derived from interfollicular epidermis are generally well-differentiated, hair follicle stem cell-derived SCCs frequently undergo EMT, efficiently form secondary tumors, and possess increased metastatic potential. These differences are due, in part, to distinct chromatin landscapes and gene regulatory networks that cooperate to prime EMT gene expression particularly in the hair follicle lineages [39].

It is increasingly recognized that EMT is not a binary process but a dynamic one which produces a variety of intermediate cell states. Indeed, subtypes of ovarian cancer with varying epithelial and mesenchymal characteristics and corresponding tumorigenic properties have been proposed [40]. A recent study elegantly demonstrated a variety of EMT phenotypes in vivo. By profiling tumor cells using cell surface marker expression and single-cell RNA-seq, Pastushenko et al. [41] uncovered subpopulation of cells with E, M, and hybrid EM states. Intriguingly, all subpopulations of cells exhibited similar tumor-initiating frequency as assessed by transplantation into immunocompromised mice. However, hybrid epithelial and mesenchymal cell populations show the greatest lung metastasis potential while mesenchymal cell populations in general showed higher levels of cellular plasticity and invasiveness. This report and others have also unveiled specific spatial localization of EMT cell populations in tumors. In breast cancer, specific CSC populations have been proposed for epithelial and mesenchymal regions within the same tumors. While ALDH marked epithelial CSCs, $CD24^{-}$ $CD44^{+}$ marked mesenchymal CSCs that were specifically located at the invasive front and were less proliferative compared to the epithelial CSCs [3]. Similarly, single-cell analyses of head-and-neck SCC have uncovered cells with partial EMT features that spatially localize to the leading edge of primary tumors and facilitate invasion [42].

EMT may form part of the mechanistic basis of the stochastic model where cells along the hierarchy may stochastically adopt a more mesenchymal cell state which will contribute to increasing stemness. Indeed, populations of non-CSC have been shown to spontaneously undergo EMT under appropriate conditions, acquire CSC-like cell surface markers and an enhanced capacity to seed tumors in mice [43, 44].

6 CSC Metabolism

The identification of deregulated cell metabolism as a hallmark of cancer has opened new avenues for cancer research. Increasingly, CSCs have been shown to harbor unique metabolic phenotypes that are distinct from non-CSCs.

Several reports suggest that CSCs are more glycolytic (Warburg effect) than other differentiated cancer cells in vitro and in vivo in osteosarcoma, glioblastoma, breast cancer, lung cancer, ovarian cancer, and colon cancer. Glycolysis was also found to be the preferred metabolic program in radioresistant sphere-forming cells in nasopharyngeal carcinoma [45] and $CD133^{+}CD49f^{+}$ TICs in hepatocellular carcinoma [46].

Apart from glycolysis, CSCs may exhibit increasing reliance on oxidative phosphorylation (OXPHOS) for metabolism. For example, patient-derived glioblastoma relied more on OXPHOS than their differentiated progeny. The same is true for sphere-forming and $CD133^{+}$ cells for both glioblastoma [47] and pancreatic ductal adenocarcinoma (PDAC) [48]. Importantly, these studies also suggest that OXPHOS is intricately linked to self-renewal and in vivo tumorigenic capacity of CSCs. Metabolic vulnerabilities associated with amino acid metabolism have also been described in CSCs. $CD166^{+}$ CSCs in NSCLC were found to exhibit high expression of the enzyme glycine decarboxylase (GLDC) which promotes tumorigenesis via its metabolic activity [49]. Following this study, using isogenic tumorsphere and adherent NSCLC models, Wang et al. (manuscript in review) identified MAT2A and methionine pathway as a vulnerability in the same $CD166^{+}$ NSCLC CSCs. Small-molecule chemical inhibition of MAT2A by FIDAS was effective in hampering its tumor-initiating capacity.

Altered lipid metabolism may be another hallmark of CSCs. Self-renewal in both hematopoietic stem cells and leukemia-initiating cells appears to be dependent on fatty acid oxidation (FAO) [50, 51] while inhibition of FAO with JAK/STAT3 inhibitors preferentially eliminates $CD44^{+}$ breast CSCs compared to non-CSCs [52]. A recent study shows that CD44 high metastasis-initiating cells of oral carcinoma express high levels of the CD36 fatty acid receptor and lipid metabolic genes [53].

Altogether, such metabolic vulnerabilities and presumably many more that are going to be uncovered with more sophisticated metabolite profiling techniques and flux analyses may be used as a basis for CSC-targeted therapy (reviewed in Ref. [54]).

7 Cell-Autonomous, Niche, or Both? Intrinsic Versus Extrinsic Factors that Govern CSC Behavior

Genetic and epigenetic changes that occur within the cell undoubtedly contribute to clonal expansion and tumor growth. In addition to these changes, the tumor microenvironment which consists of immune cells, stromal cells, blood vessels, and their secreted factors plays a critical role in determining CSC function and properties. Through paracrine interactions, the tumor microenvironment has been shown to initiate stem cell-like programs in cancer cells [55, 56]. The inflammatory microenvironment of the tumor may also influence CSC properties. During intestinal tumorigenesis, a bidirectional conversion between CSCs and non-CSCs can be triggered by an inflammatory stroma which is characterized by elevated

NF-κB and Wnt signaling, leading to the dedifferentiation of non-CSCs that acquire tumor-initiating capacity [57].

Tumor and microenvironment interactions are often bidirectional. There is evidence pointing to factors produced by CSCs and endothelial cells in the tumor microenvironment that can transform normal fibroblasts into cancer-associated fibroblasts (CAFs) (reviewed in Ref. [58]). In turn, CAFs, far from being a passive player in the tumor ecosystem, can promote tumor progression and induction of stemness. One such example is in pancreatic cancer where CAFs can enable tumor cells to undergo EMT through the secretion of cytokines such as Il-6 [59]. Moreover, CSCs in glioblastomas have been shown to secrete VEGF to promote the development of vasculature [60].

It is pertinent to note that the contribution of extrinsic factors could be very different in liquid and solid tumors. While there is evidence showing that the bone marrow niche [61], cytokines, and growth factors can govern the fate of leukemic stem cells [62], unlike blood, cells in epithelial tissues are in direct contact with different cell types and complex extracellular matrix throughout tumorigenesis, and it is reasonable to propose that cells originating in epithelial tumors respond more readily to these external factors. Such factors cause CSCs in epithelial tumors to be easily reprogrammed, resulting in remarkable plasticity, expression of varying cell surface markers, and different responses to different assays. As a result, there are oftentimes ambiguous interpretation of the existence and the nature of CSCs in solid tumors.

8 CSC and Implications for Cancer Therapy

The presence of CSCs has important implications for cancer therapy. Cancers that have a CSC-associated molecular signature often correlate with poor patient prognosis [63]. Interestingly, even in patients with diverse driver mutations, gene expression signatures that are specific to CSCs and normal stem cells are good prognostic markers for patient outcome [63, 64]. This shows that signaling pathways that drive stemness may be advantageous to the cell.

Tumors are known to evolve resistance mechanisms against commonly used therapeutic compounds and this resistance is observed in both the therapies that use specific molecular inhibitors or broad-spectrum compounds targeting proliferative cells. Even with sophisticated chemotherapeutic strategies coupled with ionizing radiation, a subset of tumor cells remain or develop resistance that later contributes directly to the development of new chemoresistant tumors. It is not clear if resistant cells preexist or develop at the time of therapy but it has been widely speculated that CSCs are highly quiescent to evade therapies specific to proliferative cells.

The high expression of known CSC markers such as ABC transporters that often efflux drugs such as doxorubicin and paclitaxel [65, 66] from the cell may further provide the link between CSC and chemoresistance. Conjugating nanoparticles to drugs can decrease the efflux activity of ABC transporters but there is limited evidence in support of this and a number of clinical studies focused on ABC transporters are underway [67]. ALDH, a known CSC marker in many tumors, is also known to be involved in chemoresistance but its exact mechanisms and the effect of its enzymatic activity remain to be characterized. β-catenin-mediated Wnt signaling, Notch signaling, and Bcl-2 pathways that are often associated with CSCs may also aid in chemo- and other drug resistance [68–70]. Notably, combined inhibition of Wnt signaling and Hedgehog signaling was sufficient in eliminating a $Lgr5^+$ cell population that emerged after vismogenib treatment of basal cell carcinoma [71]. It is important to note that the mechanisms of drug resistance by ABC transporters, ALDH, or other signaling pathways are different in different tumors and the efficacy of each CSC marker as a potential drug-resistant target must be studied in the context of a specific tumor type.

In the following section we review a few strategies that are currently used or considered to specifically target CSCs in cancer therapy (Fig. 3).

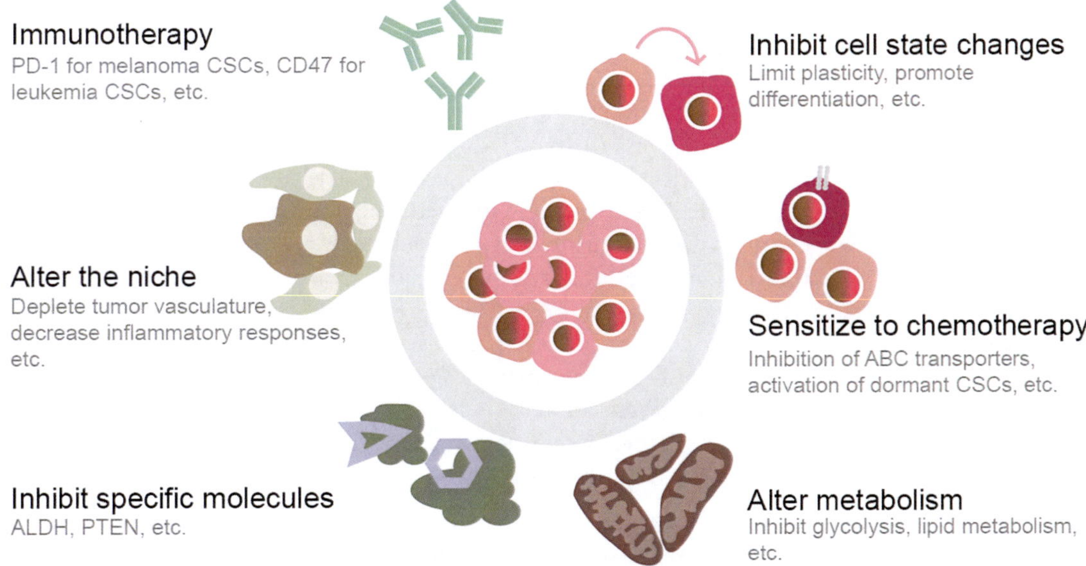

Fig. 3 Therapeutic strategies for CSCs. The unique characteristics of CSCs present new options for cancer therapy which could be complemented with existing systemic therapies such as chemotherapy (*see* text for more details)

8.1 Reversing EMT

The reversal of cell states from mesenchymal to epithelial or the induction of mesenchymal to epithelial transition (MET) is an attractive strategy to reduce cancer cell stemness, perturb cell invasion, migration and the development of metastasis. One such proof-of-concept is the activation of Protein Kinase A in mammary mesenchymal cells which was shown to be effective in inducing a more epithelial phenotype and increasing the sensitivity of cells to chemotherapy [72]. Strategies to reverse the epigenetic reprogramming induced by Zeb1 in pancreatic and breast cancer cell models have also been shown to repress stemness and overcome drug resistance [73]. Besides overcoming drug resistance, it could be envisioned that the induction of MET could limit the tumor-initiating capacity of mesenchymal cells.

A major challenge in this therapeutic strategy is to define a therapeutic window that would promote epithelialization in primary tumors while excluding metastatic cells that have already disseminated from primary tumors. The latter scenario may conceivably promote epithelial colonization leading to the undesirable formation of potentially detrimental secondary tumors.

8.2 Targeting ALDH

ALDH activity is known to be involved in chemoresistance but the exact mechanisms and the effect of its enzymatic activity remain to be characterized. Targeting ALDH could be a viable therapeutic strategy—simultaneous knockdown of ALDH and gemcitabine treatment in pancreatic adenocarcinoma cells have been shown to induce apoptosis and decrease proliferation in vitro [74]. Many small-molecule inhibitors for ALDHs have been successfully developed. Of these, inhibiting ALDH1 with diethylaminobenzaldehyde (DEAB) has been shown to sensitize $ALDH^+$ breast cells to paclitaxel and epirubicin [75]. ALDHs can also be targeted with vitamin A-related compounds known as retinoids that may increase the effectiveness of standard chemotherapy (Loo et al., manuscript in preparation) [76–78]. These drugs activate retinoid acid signaling which decreases the expression of stemness markers, promotes cellular differentiation and cell cycle arrest, decreases cellular proliferation, and reduces tumor growth in mice (reviewed in [79]).

8.3 Immunotherapy for CSCs

Immune cells could be harnessed to eradicate CSCs. In AML, targeting of the receptor CD47 by monoclonal antibodies in $CD34^+$ leukemic stem cells is sufficient to promote phagocytosis by macrophages while sparing normal cells [80]. Moreover, in combination with rituximab, anti-CD47 therapy shows an increased efficacy in eradicating non-Hodgkin lymphoma [81]. Another approach in clinical trials involves the administration of autologous dendritic cell vaccines to glioblastoma patients [82]. With increased understanding of unique molecular regulators that are unique in CSCs, CAR-T cell-based immunotherapies against

CSCs may dominate therapeutic options in the near future (reviewed in [83]).

Several challenges with this therapeutic strategy remain nonetheless. CSCs seem to have unique immune evasion features including overexpression of PD-1/PD-L1 molecules. In melanoma, the ABCB5$^+$ CSC cellular subset selectively expresses B7.2 (a CTLA4 ligand) and PD-1 [84]. Similarly, in lung SCC, Sca-1$^+$ NGFR1$^+$ cells which display increased tumor-propagating activity compared with bulk cells show enrichment for PD-L1 expression [85]. One mechanism to account for the increase in PD-L1 expression in CSCs has been proposed recently. In breast cancer, EMT may enrich PD-L1 expression in CSCs by the β-catenin/STT3/PD-L1 signaling axis. Consequently, the induction of MET downregulates PD-L1 and promotes antitumor immunity [86].

8.4 Targeting Self-Renewal and Reprogramming Pathways of CSCs

Targeting the molecular regulators of reprogramming that limits the ability of cells to gain a stem-like state or directing CSCs towards differentiation are potential therapeutic strategies for CSCs. Pharmacologic inhibition of reprogramming or self-renewal pathways in CSCs may have therapeutic value (reviewed in [87]). Several preclinical models lend support to this approach—inhibition of the Hedgehog pathway in leukemias inhibited the expansion of imatinib-resistant chronic myelod leukemia [88, 89]. Notch pathway inhibition in brain cancer promoted its sensitivity to radiation [90]. However, it is increasingly clear that combination approaches to overcome the cross talk among Notch, Hedgehog, and Wnt pathways as well as other signaling pathways would be more effective than single agents or combined single agent-chemotherapy regiment [87]. An ongoing clinical trial explores the effect of cirmtuzumab, a ROR1-based humanized monoclonal antibody drug, on patients with relapsed or refractory chronic lymphocytic leukemia [91]. Aberrant expression of ROR-1 is seen in many malignancies and has been linked to RhoGTPase activation and CSC self-renewal. However, toxicity remains a concern given that such self-renewal pathways are also activated in normal stem cells.

8.5 Targeting the CSC Niche

Some attempts to target the malicious CSC niche have already shown promise. Targeting hypoxia could be useful to eradicate quiescent and drug-resistant cells. HIF-1α and HIF-2α which promote cell cycle via c-Myc are promising targets for therapy for glioma patients [92, 93]. Antiangiogenic therapies may be helpful to limit CSC function in vivo. VEGF inhibition can deplete the tumor vasculature and ablate self-renewing CSCs and inhibit tumor growth. Similarly, blocking DLL4-mediated signaling in tumor and vascular cells is effective in inhibiting growth of colon tumor xenografts [94]. Depletion of tumor associated macrophages by inhibiting either CCR2 or M-CSF receptors resulted in decreased CSCs in pancreatic tumors, improved chemotherapeutic efficacy, inhibited metastasis, and increased antitumor T-cell responses [95].

Lastly, the disruption of chemokine receptors that are expressed on CSCs such as CXCR4 could be a viable therapeutic strategy [96, 97].

9 Summary and Perspectives

Until recently, cancer has always been perceived as a homogenous group of cells with uncontrolled proliferation and initial cancer therapies have been focused on halting cellular proliferation or promoting apoptosis. The CSC concept has provided researchers with a new way to look at tumors and rethink strategies for cancer therapy. Since the discovery of CSCs, we are now viewing and studying tumors as highly regulated models with extensive heterogeneity and clonal cell cooperations that could evolve mechanisms to undergo metastasis and resist therapy.

The presence and the biology of CSCs remain to be clarified with systematic characterization of tumor-type-specific CSCs in patients with reliable and reproducible phenotypic markers and molecular targets that are distinguishable from non-CSCs. Studies that characterize CSC content and catalog the heterogeneous cell types in patient tumors are necessary to translate laboratory findings of CSCs to potential diagnostic tools or therapeutic applications in the clinic. With the advent of single-cell techniques, researchers are now beginning to characterize the heterogeneity and the presence of CSCs in patient tumors and a refined CSC catalog in multiple tumor types is expected to be revealed in the next few years. These efforts, together with the deepening knowledge pool from various approaches will allow the design of specific therapies that target CSCs which, hopefully, may lead to the complete eradication of cancer.

References

1. Lapidot T, Sirard C, Vormoor J, Murdoch B, Hoang T, Caceres-Cortes J, Minden M, Paterson B, Caligiuri MA, Dick JE (1994) A cell initiating human acute myeloid leukaemia after transplantation into SCID mice. Nature 367(6464):645–648. https://doi.org/10.1038/367645a0
2. Bonnet D, Dick JE (1997) Human acute myeloid leukemia is organized as a hierarchy that originates from a primitive hematopoietic cell. Nat Med 3(7):730–737
3. Liu S, Cong Y, Wang D, Sun Y, Deng L, Liu Y, Martin-Trevino R, Shang L, McDermott SP, Landis MD, Hong S, Adams A, D'Angelo R, Ginestier C, Charafe-Jauffret E, Clouthier SG, Birnbaum D, Wong ST, Zhan M, Chang JC, Wicha MS (2014) Breast cancer stem cells transition between epithelial and mesenchymal states reflective of their normal counterparts. Stem Cell Reports 2 (1):78–91. https://doi.org/10.1016/j.stemcr.2013.11.009
4. de Sousa e Melo F, Kurtova AV, Harnoss JM, Kljavin N, Hoeck JD, Hung J, Anderson JE, Storm EE, Modrusan Z, Koeppen H, Dijkgraaf GJ, Piskol R, de Sauvage FJ (2017) A distinct role for Lgr5(+) stem cells in primary and metastatic colon cancer. Nature 543

(7647):676–680. https://doi.org/10.1038/nature21713

5. Barker N, van Es JH, Kuipers J, Kujala P, van den Born M, Cozijnsen M, Haegebarth A, Korving J, Begthel H, Peters PJ, Clevers H (2007) Identification of stem cells in small intestine and colon by marker gene Lgr5. Nature 449(7165):1003–1007. https://doi.org/10.1038/nature06196
6. Sanchez-Danes A, Hannezo E, Larsimont JC, Liagre M, Youssef KK, Simons BD, Blanpain C (2016) Defining the clonal dynamics leading to mouse skin tumour initiation. Nature 536(7616):298–303. https://doi.org/10.1038/nature19069
7. Krivtsov AV, Twomey D, Feng Z, Stubbs MC, Wang Y, Faber J, Levine JE, Wang J, Hahn WC, Gilliland DG, Golub TR, Armstrong SA (2006) Transformation from committed progenitor to leukaemia stem cell initiated by MLL-AF9. Nature 442(7104):818–822. https://doi.org/10.1038/nature04980
8. Lapouge G, Beck B, Nassar D, Dubois C, Dekoninck S, Blanpain C (2012) Skin squamous cell carcinoma propagating cells increase with tumour progression and invasiveness. EMBO J 31(24):4563–4575. https://doi.org/10.1038/emboj.2012.312
9. Youssef KK, Van Keymeulen A, Lapouge G, Beck B, Michaux C, Achouri Y, Sotiropoulou PA, Blanpain C (2010) Identification of the cell lineage at the origin of basal cell carcinoma. Nat Cell Biol 12(3):299–305. https://doi.org/10.1038/ncb2031
10. Kelly PN, Dakic A, Adams JM, Nutt SL, Strasser A (2007) Tumor growth need not be driven by rare cancer stem cells. Science 317 (5836):337. https://doi.org/10.1126/science.1142596
11. Boiko AD, Razorenova OV, van de Rijn M, Swetter SM, Johnson DL, Ly DP, Butler PD, Yang GP, Joshua B, Kaplan MJ, Longaker MT, Weissman IL (2010) Human melanoma-initiating cells express neural crest nerve growth factor receptor CD271. Nature 466(7302):133–137. https://doi.org/10.1038/nature09161
12. Quintana E, Shackleton M, Foster HR, Fullen DR, Sabel MS, Johnson TM, Morrison SJ (2010) Phenotypic heterogeneity among tumorigenic melanoma cells from patients that is reversible and not hierarchically organized. Cancer Cell 18(5):510–523. https://doi.org/10.1016/j.ccr.2010.10.012
13. Cheung AM, Wan TS, Leung JC, Chan LY, Huang H, Kwong YL, Liang R, Leung AY (2007) Aldehyde dehydrogenase activity in leukemic blasts defines a subgroup of acute myeloid leukemia with adverse prognosis and superior NOD/SCID engrafting potential. Leukemia 21(7):1423–1430. https://doi.org/10.1038/sj.leu.2404721
14. Ginestier C, Hur MH, Charafe-Jauffret E, Monville F, Dutcher J, Brown M, Jacquemier J, Viens P, Kleer CG, Liu S, Schott A, Hayes D, Birnbaum D, Wicha MS, Dontu G (2007) ALDH1 is a marker of normal and malignant human mammary stem cells and a predictor of poor clinical outcome. Cell Stem Cell 1(5):555–567. https://doi.org/10.1016/j.stem.2007.08.014
15. Martinez-Cruzado L, Tornin J, Santos L, Rodriguez A, Garcia-Castro J, Moris F, Rodriguez R (2016) Aldh1 expression and activity increase during tumor evolution in sarcoma cancer stem cell populations. Sci Rep 6:27878. https://doi.org/10.1038/srep27878
16. Dontu G, Abdallah WM, Foley JM, Jackson KW, Clarke MF, Kawamura MJ, Wicha MS (2003) In vitro propagation and transcriptional profiling of human mammary stem/progenitor cells. Genes Dev 17 (10):1253–1270. https://doi.org/10.1101/gad.1061803
17. Stewart JM, Shaw PA, Gedye C, Bernardini MQ, Neel BG, Ailles LE (2011) Phenotypic heterogeneity and instability of human ovarian tumor-initiating cells. Proc Natl Acad Sci U S A 108(16):6468–6473. https://doi.org/10.1073/pnas.1005529108
18. Shackleton M, Vaillant F, Simpson KJ, Stingl J, Smyth GK, Asselin-Labat ML, Wu L, Lindeman GJ, Visvader JE (2006) Generation of a functional mammary gland from a single stem cell. Nature 439 (7072):84–88. https://doi.org/10.1038/nature04372
19. Stingl J, Eirew P, Ricketson I, Shackleton M, Vaillant F, Choi D, Li HI, Eaves CJ (2006) Purification and unique properties of mammary epithelial stem cells. Nature 439 (7079):993–997. https://doi.org/10.1038/nature04496
20. Van Keymeulen A, Rocha AS, Ousset M, Beck B, Bouvencourt G, Rock J, Sharma N, Dekoninck S, Blanpain C (2011) Distinct stem cells contribute to mammary gland development and maintenance. Nature 479 (7372):189–193. https://doi.org/10.1038/nature10573
21. Lilja AM, Rodilla V, Huyghe M, Hannezo E, Landragin C, Renaud O, Leroy O, Rulands S, Simons BD, Fre S (2018) Clonal analysis of Notch1-expressing cells reveals the existence of unipotent stem cells that retain long-term

plasticity in the embryonic mammary gland. Nat Cell Biol 20(6):677–687. https://doi.org/10.1038/s41556-018-0108-1

22. Van Keymeulen A, Lee MY, Ousset M, Brohee S, Rorive S, Giraddi RR, Wuidart A, Bouvencourt G, Dubois C, Salmon I, Sotiriou C, Phillips WA, Blanpain C (2015) Reactivation of multipotency by oncogenic PIK3CA induces breast tumour heterogeneity. Nature 525(7567):119–123. https://doi.org/10.1038/nature14665
23. Boumahdi S, Driessens G, Lapouge G, Rorive S, Nassar D, Le Mercier M, Delatte B, Caauwe A, Lenglez S, Nkusi E, Brohee S, Salmon I, Dubois C, del Marmol V, Fuks F, Beck B, Blanpain C (2014) SOX2 controls tumour initiation and cancer stem-cell functions in squamous-cell carcinoma. Nature 511(7508):246–250. https://doi.org/10.1038/nature13305
24. Chen J, Li Y, Yu TS, McKay RM, Burns DK, Kernie SG, Parada LF (2012) A restricted cell population propagates glioblastoma growth after chemotherapy. Nature 488 (7412):522–526. https://doi.org/10.1038/nature11287
25. Zomer A, Ellenbroek SI, Ritsma L, Beerling E, Vrisekoop N, Van Rheenen J (2013) Intravital imaging of cancer stem cell plasticity in mammary tumors. Stem Cells 31 (3):602–606. https://doi.org/10.1002/stem.1296
26. Kozar S, Morrissey E, Nicholson AM, van der Heijden M, Zecchini HI, Kemp R, Tavare S, Vermeulen L, Winton DJ (2013) Continuous clonal labeling reveals small numbers of functional stem cells in intestinal crypts and adenomas. Cell Stem Cell 13(5):626–633. https://doi.org/10.1016/j.stem.2013.08.001
27. Humphries A, Cereser B, Gay LJ, Miller DS, Das B, Gutteridge A, Elia G, Nye E, Jeffery R, Poulsom R, Novelli MR, Rodriguez-Justo M, McDonald SA, Wright NA, Graham TA (2013) Lineage tracing reveals multipotent stem cells maintain human adenomas and the pattern of clonal expansion in tumor evolution. Proc Natl Acad Sci U S A 110(27): E2490–E2499. https://doi.org/10.1073/pnas.1220353110
28. Chang HH, Hemberg M, Barahona M, Ingber DE, Huang S (2008) Transcriptome-wide noise controls lineage choice in mammalian progenitor cells. Nature 453 (7194):544–547. https://doi.org/10.1038/nature06965
29. Patel AP, Tirosh I, Trombetta JJ, Shalek AK, Gillespie SM, Wakimoto H, Cahill DP, Nahed BV, Curry WT, Martuza RL, Louis DN, Rozenblatt-Rosen O, Suva ML, Regev A, Bernstein BE (2014) Single-cell RNA-seq highlights intratumoral heterogeneity in primary glioblastoma. Science 344 (6190):1396–1401. https://doi.org/10.1126/science.1254257
30. Li H, Courtois ET, Sengupta D, Tan Y, Chen KH, Goh JJL, Kong SL, Chua C, Hon LK, Tan WS, Wong M, Choi PJ, Wee LJK, Hillmer AM, Tan IB, Robson P, Prabhakar S (2017) Reference component analysis of single-cell transcriptomes elucidates cellular heterogeneity in human colorectal tumors. Nat Genet 49 (5):708–718. https://doi.org/10.1038/ng.3818
31. Wahl GM, Spike BT (2017) Cell state plasticity, stem cells, EMT, and the generation of intra-tumoral heterogeneity. NPJ Breast Cancer 3:14. https://doi.org/10.1038/s41523-017-0012-z
32. Giraddi RR, Chung CY, Heinz RE, Balcioglu O, Novotny M, Trejo CL, Dravis C, Hagos BM, Mehrabad EM, Rodewald LW, Hwang JY, Fan C, Lasken R, Varley KE, Perou CM, Wahl GM, Spike BT (2018) Single-cell transcriptomes distinguish stem cell state changes and lineage specification programs in early mammary gland development. Cell Rep 24(6):1653–1666 e1657. https://doi.org/10.1016/j.celrep.2018.07.025
33. Spike BT, Engle DD, Lin JC, Cheung SK, La J, Wahl GM (2012) A mammary stem cell population identified and characterized in late embryogenesis reveals similarities to human breast cancer. Cell Stem Cell 10 (2):183–197. https://doi.org/10.1016/j.stem.2011.12.018
34. Mani SA, Guo W, Liao MJ, Eaton EN, Ayyanan A, Zhou AY, Brooks M, Reinhard F, Zhang CC, Shipitsin M, Campbell LL, Polyak K, Brisken C, Yang J, Weinberg RA (2008) The epithelial-mesenchymal transition generates cells with properties of stem cells. Cell 133(4):704–715. https://doi.org/10.1016/j.cell.2008.03.027
35. Morel AP, Lievre M, Thomas C, Hinkal G, Ansieau S, Puisieux A (2008) Generation of breast cancer stem cells through epithelial-mesenchymal transition. PLoS One 3(8): e2888. https://doi.org/10.1371/journal.pone.0002888
36. Tam WL, Lu H, Buikhuisen J, Soh BS, Lim E, Reinhardt F, Wu ZJ, Krall JA, Bierie B, Guo W, Chen X, Liu XS, Brown M, Lim B, Weinberg RA (2013) Protein kinase C alpha is a central signaling node and therapeutic target

for breast cancer stem cells. Cancer Cell 24 (3):347–364. https://doi.org/10.1016/j.ccr.2013.08.005

37. Krebs AM, Mitschke J, Lasierra Losada M, Schmalhofer O, Boerries M, Busch H, Boettcher M, Mougiakakos D, Reichardt W, Bronsert P, Brunton VG, Pilarsky C, Winkler TH, Brabletz S, Stemmler MP, Brabletz T (2017) The EMT-activator Zeb1 is a key factor for cell plasticity and promotes metastasis in pancreatic cancer. Nat Cell Biol 19 (5):518–529. https://doi.org/10.1038/ncb3513
38. Guo W, Keckesova Z, Donaher JL, Shibue T, Tischler V, Reinhardt F, Itzkovitz S, Noske A, Zurrer-Hardi U, Bell G, Tam WL, Mani SA, van Oudenaarden A, Weinberg RA (2012) Slug and Sox9 cooperatively determine the mammary stem cell state. Cell 148 (5):1015–1028. https://doi.org/10.1016/j.cell.2012.02.008
39. Latil M, Nassar D, Beck B, Boumahdi S, Wang L, Brisebarre A, Dubois C, Nkusi E, Lenglez S, Checinska A, Vercauteren Drubbel A, Devos M, Declercq W, Yi R, Blanpain C (2017) Cell-type-specific chromatin states differentially prime squamous cell carcinoma tumor-initiating cells for epithelial to mesenchymal transition. Cell Stem Cell 20 (2):191–204.e5. https://doi.org/10.1016/j.stem.2016.10.018
40. Tan TZ, Miow QH, Huang RY, Wong MK, Ye J, Lau JA, Wu MC, Bin Abdul Hadi LH, Soong R, Choolani M, Davidson B, Nesland JM, Wang LZ, Matsumura N, Mandai M, Konishi I, Goh BC, Chang JT, Thiery JP, Mori S (2013) Functional genomics identifies five distinct molecular subtypes with clinical relevance and pathways for growth control in epithelial ovarian cancer. EMBO Mol Med 5 (7):1051–1066. https://doi.org/10.1002/emmm.201201823
41. Pastushenko I, Brisebarre A, Sifrim A, Fioramonti M, Revenco T, Boumahdi S, Van Keymeulen A, Brown D, Moers V, Lemaire S, De Clercq S, Minguijon E, Balsat C, Sokolow Y, Dubois C, De Cock F, Scozzaro S, Sopena F, Lanas A, D'Haene N, Salmon I, Marine JC, Voet T, Sotiropoulou PA, Blanpain C (2018) Identification of the tumour transition states occurring during EMT. Nature 556(7702):463–468. https://doi.org/10.1038/s41586-018-0040-3
42. Puram SV, Tirosh I, Parikh AS, Patel AP, Yizhak K, Gillespie S, Rodman C, Luo CL, Mroz EA, Emerick KS, Deschler DG, Varvares MA, Mylvaganam R, Rozenblatt-Rosen O, Rocco JW, Faquin WC, Lin DT, Regev A, Bernstein BE (2017) Single-cell transcriptomic analysis of primary and metastatic tumor ecosystems in head and neck cancer. Cell 171(7):1611–1624. https://doi.org/10.1016/j.cell.2017.10.044
43. Chaffer CL, Brueckmann I, Scheel C, Kaestli AJ, Wiggins PA, Rodrigues LO, Brooks M, Reinhardt F, Su Y, Polyak K, Arendt LM, Kuperwasser C, Bierie B, Weinberg RA (2011) Normal and neoplastic nonstem cells can spontaneously convert to a stem-like state. Proc Natl Acad Sci U S A 108 (19):7950–7955. https://doi.org/10.1073/pnas.1102454108
44. Chaffer CL, Marjanovic ND, Lee T, Bell G, Kleer CG, Reinhardt F, D'Alessio AC, Young RA, Weinberg RA (2013) Poised chromatin at the ZEB1 promoter enables breast cancer cell plasticity and enhances tumorigenicity. Cell 154(1):61–74. https://doi.org/10.1016/j.cell.2013.06.005
45. Shen YA, Wang CY, Hsieh YT, Chen YJ, Wei YH (2015) Metabolic reprogramming orchestrates cancer stem cell properties in nasopharyngeal carcinoma. Cell Cycle 14 (1):86–98. https://doi.org/10.4161/15384101.2014.974419
46. Chen Z, Wang Z, Guo W, Zhang Z, Zhao F, Zhao Y, Jia D, Ding J, Wang H, Yao M, He X (2015) TRIM35 interacts with pyruvate kinase isoform M2 to suppress the Warburg effect and tumorigenicity in hepatocellular carcinoma. Oncogene 34(30):3946–3956. https://doi.org/10.1038/onc.2014.325
47. Janiszewska M, Suva ML, Riggi N, Houtkooper RH, Auwerx J, Clement-Schatlo V, Radovanovic I, Rheinbay E, Provero P, Stamenkovic I (2012) Imp2 controls oxidative phosphorylation and is crucial for preserving glioblastoma cancer stem cells. Genes Dev 26 (17):1926–1944. https://doi.org/10.1101/gad.188292.112
48. Sancho P, Burgos-Ramos E, Tavera A, Bou Kheir T, Jagust P, Schoenhals M, Barneda D, Sellers K, Campos-Olivas R, Grana O, Viera CR, Yuneva M, Sainz B Jr, Heeschen C (2015) MYC/PGC-1alpha balance determines the metabolic phenotype and plasticity of pancreatic cancer stem cells. Cell Metab 22 (4):590–605. https://doi.org/10.1016/j.cmet.2015.08.015
49. Zhang WC, Shyh-Chang N, Yang H, Rai A, Umashankar S, Ma S, Soh BS, Sun LL, Tai BC, Nga ME, Bhakoo KK, Jayapal SR, Nichane M, Yu Q, Ahmed DA, Tan C, Sing WP, Tam J, Thirugananam A, Noghabi MS, Pang YH, Ang HS, Mitchell W, Robson P, Kaldis P, Soo RA, Swarup S, Lim EH, Lim B

(2012) Glycine decarboxylase activity drives non-small cell lung cancer tumor-initiating cells and tumorigenesis. Cell 148 (1–2):259–272. https://doi.org/10.1016/j.cell.2011.11.050

50. Samudio I, Harmancey R, Fiegl M, Kantarjian H, Konopleva M, Korchin B, Kaluarachchi K, Bornmann W, Duvvuri S, Taegtmeyer H, Andreeff M (2010) Pharmacologic inhibition of fatty acid oxidation sensitizes human leukemia cells to apoptosis induction. J Clin Invest 120(1):142–156. https://doi.org/10.1172/JCI38942
51. Ito K, Carracedo A, Weiss D, Arai F, Ala U, Avigan DE, Schafer ZT, Evans RM, Suda T, Lee CH, Pandolfi PP (2012) A PML-PPAR-delta pathway for fatty acid oxidation regulates hematopoietic stem cell maintenance. Nat Med 18(9):1350–1358. https://doi.org/10.1038/nm.2882
52. Wang T, Fahrmann JF, Lee H, Li YJ, Tripathi SC, Yue C, Zhang C, Lifshitz V, Song J, Yuan Y, Somlo G, Jandial R, Ann D, Hanash S, Jove R, Yu H (2018) JAK/STAT3-regulated fatty acid beta-oxidation is critical for breast cancer stem cell self-renewal and chemoresistance. Cell Metab 27 (1):136–150 e135. https://doi.org/10.1016/j.cmet.2017.11.001
53. Pascual G, Avgustinova A, Mejetta S, Martin M, Castellanos A, Attolini CS, Berenguer A, Prats N, Toll A, Hueto JA, Bescos C, Di Croce L, Benitah SA (2017) Targeting metastasis-initiating cells through the fatty acid receptor CD36. Nature 541 (7635):41–45. https://doi.org/10.1038/nature20791
54. De Francesco EM, Sotgia F, Lisanti MP (2018) Cancer stem cells (CSCs): metabolic strategies for their identification and eradication. Biochem J 475(9):1611–1634. https://doi.org/10.1042/BCJ20170164
55. Charles N, Ozawa T, Squatrito M, Bleau AM, Brennan CW, Hambardzumyan D, Holland EC (2010) Perivascular nitric oxide activates notch signaling and promotes stem-like character in PDGF-induced glioma cells. Cell Stem Cell 6(2):141–152. https://doi.org/10.1016/j.stem.2010.01.001
56. Vermeulen L, De Sousa EMF, van der Heijden M, Cameron K, de Jong JH, Borovski T, Tuynman JB, Todaro M, Merz C, Rodermond H, Sprick MR, Kemper K, Richel DJ, Stassi G, Medema JP (2010) Wnt activity defines colon cancer stem cells and is regulated by the microenvironment. Nat Cell Biol 12(5):468–476. https://doi.org/10.1038/ncb2048
57. Schwitalla S, Fingerle AA, Cammareri P, Nebelsiek T, Goktuna SI, Ziegler PK, Canli O, Heijmans J, Huels DJ, Moreaux G, Rupec RA, Gerhard M, Schmid R, Barker N, Clevers H, Lang R, Neumann J, Kirchner T, Taketo MM, van den Brink GR, Sansom OJ, Arkan MC, Greten FR (2013) Intestinal tumorigenesis initiated by dedifferentiation and acquisition of stem-cell-like properties. Cell 152(1–2):25–38. https://doi.org/10.1016/j.cell.2012.12.012
58. Kalluri R, Zeisberg M (2006) Fibroblasts in cancer. Nat Rev Cancer 6(5):392–401. https://doi.org/10.1038/nrc1877
59. Guan J, Zhang H, Wen Z, Gu Y, Cheng Y, Sun Y, Zhang T, Jia C, Lu Z, Chen J (2014) Retinoic acid inhibits pancreatic cancer cell migration and EMT through the downregulation of IL-6 in cancer associated fibroblast cells. Cancer Lett 345(1):132–139. https://doi.org/10.1016/j.canlet.2013.12.006
60. Gilbertson RJ, Rich JN (2007) Making a tumour's bed: glioblastoma stem cells and the vascular niche. Nat Rev Cancer 7 (10):733–736. https://doi.org/10.1038/nrc2246
61. Zhang B, Nguyen LXT, Li L, Zhao D, Kumar B, Wu H, Lin A, Pellicano F, Hopcroft L, Su YL, Copland M, Holyoake TL, Kuo CJ, Bhatia R, Snyder DS, Ali H, Stein AS, Brewer C, Wang H, McDonald T, Swiderski P, Troadec E, Chen CC, Dorrance A, Pullarkat V, Yuan YC, Perrotti D, Carlesso N, Forman SJ, Kortylewski M, Kuo YH, Marcucci G (2018) Bone marrow niche trafficking of miR-126 controls the self-renewal of leukemia stem cells in chronic myelogenous leukemia. Nat Med 24(4):450–462. https://doi.org/10.1038/nm.4499
62. Wei J, Wunderlich M, Fox C, Alvarez S, Cigudosa JC, Wilhelm JS, Zheng Y, Cancelas JA, Gu Y, Jansen M, Dimartino JF, Mulloy JC (2008) Microenvironment determines lineage fate in a human model of MLL-AF9 leukemia. Cancer Cell 13(6):483–495. https://doi.org/10.1016/j.ccr.2008.04.020
63. Eppert K, Takenaka K, Lechman ER, Waldron L, Nilsson B, van Galen P, Metzeler KH, Poeppl A, Ling V, Beyene J, Canty AJ, Danska JS, Bohlander SK, Buske C, Minden MD, Golub TR, Jurisica I, Ebert BL, Dick JE (2011) Stem cell gene expression programs influence clinical outcome in human leukemia. Nat Med 17(9):1086–1093. https://doi.org/10.1038/nm.2415
64. Bartholdy B, Christopeit M, Will B, Mo Y, Barreyro L, Yu Y, Bhagat TD, Okoye-Okafor

UC, Todorova TI, Greally JM, Levine RL, Melnick A, Verma A, Steidl U (2014) HSC commitment-associated epigenetic signature is prognostic in acute myeloid leukemia. J Clin Invest 124(3):1158–1167. https://doi.org/10.1172/JCI71264

65. Chow EK, Fan LL, Chen X, Bishop JM (2012) Oncogene-specific formation of chemoresistant murine hepatic cancer stem cells. Hepatology 56(4):1331–1341. https://doi.org/10.1002/hep.25776
66. Litman T, Brangi M, Hudson E, Fetsch P, Abati A, Ross DD, Miyake K, Resau JH, Bates SE (2000) The multidrug-resistant phenotype associated with overexpression of the new ABC half-transporter, MXR (ABCG2). J Cell Sci 113(Pt 11):2011–2021
67. Bugde P, Biswas R, Merien F, Lu J, Liu DX, Chen M, Zhou S, Li Y (2017) The therapeutic potential of targeting ABC transporters to combat multi-drug resistance. Expert Opin Ther Targets 21(5):511–530. https://doi.org/10.1080/14728222.2017.1310841
68. Ma S, Lee TK, Zheng BJ, Chan KW, Guan XY (2008) CD133+ HCC cancer stem cells confer chemoresistance by preferential expression of the Akt/PKB survival pathway. Oncogene 27(12):1749–1758. https://doi.org/10.1038/sj.onc.1210811
69. Meng RD, Shelton CC, Li YM, Qin LX, Notterman D, Paty PB, Schwartz GK (2009) Gamma-Secretase inhibitors abrogate oxaliplatin-induced activation of the Notch-1 signaling pathway in colon cancer cells resulting in enhanced chemosensitivity. Cancer Res 69(2):573–582. https://doi.org/10.1158/0008-5472.CAN-08-2088
70. Yang W, Yan HX, Chen L, Liu Q, He YQ, Yu LX, Zhang SH, Huang DD, Tang L, Kong XN, Chen C, Liu SQ, Wu MC, Wang HY (2008) Wnt/beta-catenin signaling contributes to activation of normal and tumorigenic liver progenitor cells. Cancer Res 68 (11):4287–4295. https://doi.org/10.1158/0008-5472.CAN-07-6691
71. Sanchez-Danes A, Larsimont JC, Liagre M, Munoz-Couselo E, Lapouge G, Brisebarre A, Dubois C, Suppa M, Sukumaran V, Del Marmol V, Tabernero J, Blanpain C (2018) A slow-cycling LGR5 tumour population mediates basal cell carcinoma relapse after therapy. Nature 562 (7727):434–438. https://doi.org/10.1038/s41586-018-0603-3
72. Pattabiraman DR, Bierie B, Kober KI, Thiru P, Krall JA, Zill C, Reinhardt F, Tam WL, Weinberg RA (2016) Activation of PKA leads to mesenchymal-to-epithelial transition and loss of tumor-initiating ability. Science 351(6277):aad3680. https://doi.org/10.1126/science.aad3680
73. Meidhof S, Brabletz S, Lehmann W, Preca BT, Mock K, Ruh M, Schuler J, Berthold M, Weber A, Burk U, Lubbert M, Puhr M, Culig Z, Wellner U, Keck T, Bronsert P, Kusters S, Hopt UT, Stemmler MP, Brabletz T (2015) ZEB1-associated drug resistance in cancer cells is reversed by the class I HDAC inhibitor mocetinostat. EMBO Mol Med 7 (6):831–847. https://doi.org/10.15252/emmm.201404396
74. Duong HQ, Hwang JS, Kim HJ, Kang HJ, Seong YS, Bae I (2012) Aldehyde dehydrogenase 1A1 confers intrinsic and acquired resistance to gemcitabine in human pancreatic adenocarcinoma MIA PaCa-2 cells. Int J Oncol 41(3):855–861. https://doi.org/10.3892/ijo.2012.1516
75. Croker AK, Allan AL (2012) Inhibition of aldehyde dehydrogenase (ALDH) activity reduces chemotherapy and radiation resistance of stem-like ALDHhiCD44(+) human breast cancer cells. Breast Cancer Res Treat 133(1):75–87. https://doi.org/10.1007/s10549-011-1692-y
76. Formelli F, Cleris L (1993) Synthetic retinoid fenretinide is effective against a human ovarian carcinoma xenograft and potentiates cisplatin activity. Cancer Res 53(22):5374–5376
77. Shalinsky DR, Bischoff ED, Gregory ML, Lamph WW, Heyman RA, Hayes JS, Thomazy V, Davies PJ (1996) Enhanced antitumor efficacy of cisplatin in combination with ALRT1057 (9-cis retinoic acid) in human oral squamous carcinoma xenografts in nude mice. Clin Cancer Res 2(3):511–520
78. Pettersson F, Colston KW, Dalgleish AG (2001) Retinoic acid enhances the cytotoxic effects of gemcitabine and cisplatin in pancreatic adenocarcinoma cells. Pancreas 23 (3):273–279
79. Clark DW, Palle K (2016) Aldehyde dehydrogenases in cancer stem cells: potential as therapeutic targets. Ann Transl Med 4(24):518. https://doi.org/10.21037/atm.2016.11.82
80. Majeti R, Chao MP, Alizadeh AA, Pang WW, Jaiswal S, Gibbs KD Jr, van Rooijen N, Weissman IL (2009) CD47 is an adverse prognostic factor and therapeutic antibody target on human acute myeloid leukemia stem cells. Cell 138(2):286–299. https://doi.org/10.1016/j.cell.2009.05.045
81. Chao MP, Alizadeh AA, Tang C, Myklebust JH, Varghese B, Gill S, Jan M, Cha AC, Chan CK, Tan BT, Park CY, Zhao F, Kohrt HE, Malumbres R, Briones J, Gascoyne RD,

Lossos IS, Levy R, Weissman IL, Majeti R (2010) Anti-CD47 antibody synergizes with rituximab to promote phagocytosis and eradicate non-Hodgkin lymphoma. Cell 142 (5):699–713. https://doi.org/10.1016/j.cell.2010.07.044

82. Vik-Mo EO, Nyakas M, Mikkelsen BV, Moe MC, Due-Tonnesen P, Suso EM, Saeboe-Larssen S, Sandberg C, Brinchmann JE, Helseth E, Rasmussen AM, Lote K, Aamdal S, Gaudernack G, Kvalheim G, Langmoen IA (2013) Therapeutic vaccination against autologous cancer stem cells with mRNA-transfected dendritic cells in patients with glioblastoma. Cancer Immunol Immunother 62(9):1499–1509. https://doi.org/10.1007/s00262-013-1453-3

83. Guo Y, Feng K, Wang Y, Han W (2018) Targeting cancer stem cells by using chimeric antigen receptor-modified T cells: a potential and curable approach for cancer treatment. Protein Cell 9(6):516–526. https://doi.org/10.1007/s13238-017-0394-6

84. Schatton T, Schutte U, Frank NY, Zhan Q, Hoerning A, Robles SC, Zhou J, Hodi FS, Spagnoli GC, Murphy GF, Frank MH (2010) Modulation of T-cell activation by malignant melanoma initiating cells. Cancer Res 70 (2):697–708. https://doi.org/10.1158/0008-5472.CAN-09-1592

85. Xu C, Fillmore CM, Koyama S, Wu H, Zhao Y, Chen Z, Herter-Sprie GS, Akbay EA, Tchaicha JH, Altabef A, Reibel JB, Walton Z, Ji H, Watanabe H, Janne PA, Castrillon DH, Rustgi AK, Bass AJ, Freeman GJ, Padera RF, Dranoff G, Hammerman PS, Kim CF, Wong KK (2014) Loss of Lkb1 and Pten leads to lung squamous cell carcinoma with elevated PD-L1 expression. Cancer Cell 25 (5):590–604. https://doi.org/10.1016/j.ccr.2014.03.033

86. Hsu JM, Xia W, Hsu YH, Chan LC, Yu WH, Cha JH, Chen CT, Liao HW, Kuo CW, Khoo KH, Hsu JL, Li CW, Lim SO, Chang SS, Chen YC, Ren GX, Hung MC (2018) STT3-dependent PD-L1 accumulation on cancer stem cells promotes immune evasion. Nat Commun 9(1):1908. https://doi.org/10.1038/s41467-018-04313-6

87. Takebe N, Miele L, Harris PJ, Jeong W, Bando H, Kahn M, Yang SX, Ivy SP (2015) Targeting notch, hedgehog, and Wnt pathways in cancer stem cells: clinical update. Nat Rev Clin Oncol 12(8):445–464. https://doi.org/10.1038/nrclinonc.2015.61

88. Dierks C, Beigi R, Guo GR, Zirlik K, Stegert MR, Manley P, Trussell C, Schmitt-Graeff A, Landwerlin K, Veelken H, Warmuth M (2008) Expansion of Bcr-Abl-positive leukemic stem cells is dependent on hedgehog pathway activation. Cancer Cell 14 (3):238–249. https://doi.org/10.1016/j.ccr.2008.08.003

89. Zhao C, Chen A, Jamieson CH, Fereshteh M, Abrahamsson A, Blum J, Kwon HY, Kim J, Chute JP, Rizzieri D, Munchhof M, VanArsdale T, Beachy PA, Reya T (2009) Hedgehog signalling is essential for maintenance of cancer stem cells in myeloid leukaemia. Nature 458(7239):776–779. https://doi.org/10.1038/nature07737

90. Wang J, Wakeman TP, Lathia JD, Hjelmeland AB, Wang XF, White RR, Rich JN, Sullenger BA (2010) Notch promotes radioresistance of glioma stem cells. Stem Cells 28(1):17–28. https://doi.org/10.1002/stem.261

91. Choi MY, Widhopf GF 2nd, Ghia EM, Kidwell RL, Hasan MK, Yu J, Rassenti LZ, Chen L, Chen Y, Pittman E, Pu M, Messer K, Prussak CE, Castro JE, Jamieson C, Kipps TJ (2018) Phase I trial: Cirmtuzumab inhibits ROR1 Signaling and Stemness signatures in patients with chronic lymphocytic Leukemia. Cell Stem Cell 22 (6):951–959 e953. https://doi.org/10.1016/j.stem.2018.05.018

92. Gordan JD, Bertout JA, Hu CJ, Diehl JA, Simon MC (2007) HIF-2alpha promotes hypoxic cell proliferation by enhancing c-myc transcriptional activity. Cancer Cell 11 (4):335–347. https://doi.org/10.1016/j.ccr.2007.02.006

93. Li Z, Bao S, Wu Q, Wang H, Eyler C, Sathornsumetee S, Shi Q, Cao Y, Lathia J, McLendon RE, Hjelmeland AB, Rich JN (2009) Hypoxia-inducible factors regulate tumorigenic capacity of glioma stem cells. Cancer Cell 15(6):501–513. https://doi.org/10.1016/j.ccr.2009.03.018

94. Hoey T, Yen WC, Axelrod F, Basi J, Donigian L, Dylla S, Fitch-Bruhns M, Lazetic S, Park IK, Sato A, Satyal S, Wang X, Clarke MF, Lewicki J, Gurney A (2009) DLL4 blockade inhibits tumor growth and reduces tumor-initiating cell frequency. Cell Stem Cell 5(2):168–177. https://doi.org/10.1016/j.stem.2009.05.019

95. Mitchem JB, Brennan DJ, Knolhoff BL, Belt BA, Zhu Y, Sanford DE, Belaygorod L, Carpenter D, Collins L, Piwnica-Worms D, Hewitt S, Udupi GM, Gallagher WM, Wegner C, West BL, Wang-Gillam A, Goedegebuure P, Linehan DC, DeNardo DG (2013) Targeting tumor-infiltrating macrophages decreases tumor-initiating cells, relieves immunosuppression, and improves

chemotherapeutic responses. Cancer Res 73 (3):1128–1141. https://doi.org/10.1158/0008-5472.CAN-12-2731

96. Nagasawa T, Hirota S, Tachibana K, Takakura N, Nishikawa S, Kitamura Y, Yoshida N, Kikutani H, Kishimoto T (1996) Defects of B-cell lymphopoiesis and bone-marrow myelopoiesis in mice lacking the CXC chemokine PBSF/SDF-1. Nature 382 (6592):635–638. https://doi.org/10.1038/382635a0
97. Ma Q, Jones D, Borghesani PR, Segal RA, Nagasawa T, Kishimoto T, Bronson RT, Springer TA (1998) Impaired B-lymphopoiesis, myelopoiesis, and derailed cerebellar neuron migration in CXCR4- and SDF-1-deficient mice. Proc Natl Acad Sci U S A 95(16):9448–9453
98. van Rhenen A, van Dongen GA, Kelder A, Rombouts EJ, Feller N, Moshaver B, Stigter-van Walsum M, Zweegman S, Ossenkoppele GJ, Jan Schuurhuis G (2007) The novel AML stem cell associated antigen CLL-1 aids in discrimination between normal and leukemic stem cells. Blood 110(7):2659–2666. https://doi.org/10.1182/blood-2007-03-083048
99. Hosen N, Park CY, Tatsumi N, Oji Y, Sugiyama H, Gramatzki M, Krensky AM, Weissman IL (2007) CD96 is a leukemic stem cell-specific marker in human acute myeloid leukemia. Proc Natl Acad Sci U S A 104 (26):11008–11013. https://doi.org/10.1073/pnas.0704271104
100. Jan M, Chao MP, Cha AC, Alizadeh AA, Gentles AJ, Weissman IL, Majeti R (2011) Prospective separation of normal and leukemic stem cells based on differential expression of TIM3, a human acute myeloid leukemia stem cell marker. Proc Natl Acad Sci U S A 108(12):5009–5014. https://doi.org/10.1073/pnas.1100551108
101. Saito Y, Kitamura H, Hijikata A, Tomizawa-Murasawa M, Tanaka S, Takagi S, Uchida N, Suzuki N, Sone A, Najima Y, Ozawa H, Wake A, Taniguchi S, Shultz LD, Ohara O, Ishikawa F (2010) Identification of therapeutic targets for quiescent, chemotherapy-resistant human leukemia stem cells. Sci Transl Med 2(17):17ra19. https://doi.org/10.1126/scitranslmed.3000349
102. Yang YM, Chang JW (2008) Bladder cancer initiating cells (BCICs) are among EMA-CD44v6+ subset: novel methods for isolating undetermined cancer stem (initiating) cells. Cancer Investig 26(7):725–733. https://doi.org/10.1080/07357900801941845
103. Wang L, Park P, Lin CY (2009) Characterization of stem cell attributes in human osteosarcoma cell lines. Cancer Biol Ther 8 (6):543–552. https://doi.org/10.4161/cbt.8.6.7695
104. Al-Hajj M, Wicha MS, Benito-Hernandez A, Morrison SJ, Clarke MF (2003) Prospective identification of tumorigenic breast cancer cells. Proc Natl Acad Sci U S A 100 (7):3983–3988. https://doi.org/10.1073/pnas.0530291100
105. Singh SK, Hawkins C, Clarke ID, Squire JA, Bayani J, Hide T, Henkelman RM, Cusimano MD, Dirks PB (2004) Identification of human brain tumour initiating cells. Nature 432(7015):396–401. https://doi.org/10.1038/nature03128
106. Ricci-Vitiani L, Lombardi DG, Pilozzi E, Biffoni M, Todaro M, Peschle C, De Maria R (2007) Identification and expansion of human colon-cancer-initiating cells. Nature 445(7123):111–115. https://doi.org/10.1038/nature05384
107. O'Brien CA, Pollett A, Gallinger S, Dick JE (2007) A human colon cancer cell capable of initiating tumour growth in immunodeficient mice. Nature 445(7123):106–110. https://doi.org/10.1038/nature05372
108. Shimokawa M, Ohta Y, Nishikori S, Matano M, Takano A, Fujii M, Date S, Sugimoto S, Kanai T, Sato T (2017) Visualization and targeting of LGR5(+) human colon cancer stem cells. Nature 545 (7653):187–192. https://doi.org/10.1038/nature22081
109. Shi C, Tian R, Wang M, Wang X, Jiang J, Zhang Z, Li X, He Z, Gong W, Qin R (2010) CD44+ CD133+ population exhibits cancer stem cell-like characteristics in human gallbladder carcinoma. Cancer Biol Ther 10 (11):1182–1190
110. Takaishi S, Okumura T, Tu S, Wang SS, Shibata W, Vigneshwaran R, Gordon SA, Shimada Y, Wang TC (2009) Identification of gastric cancer stem cells using the cell surface marker CD44. Stem Cells 27 (5):1006–1020. https://doi.org/10.1002/stem.30
111. Zhang C, Li C, He F, Cai Y, Yang H (2011) Identification of CD44+CD24+ gastric cancer stem cells. J Cancer Res Clin Oncol 137 (11):1679–1686. https://doi.org/10.1007/s00432-011-1038-5
112. Prince ME, Sivanandan R, Kaczorowski A, Wolf GT, Kaplan MJ, Dalerba P, Weissman IL, Clarke MF, Ailles LE (2007) Identification of a subpopulation of cells with cancer

stem cell properties in head and neck squamous cell carcinoma. Proc Natl Acad Sci U S A 104(3):973–978. https://doi.org/10.1073/pnas.0610117104

113. Han J, Fujisawa T, Husain SR, Puri RK (2014) Identification and characterization of cancer stem cells in human head and neck squamous cell carcinoma. BMC Cancer 14:173. https://doi.org/10.1186/1471-2407-14-173

114. Yin S, Li J, Hu C, Chen X, Yao M, Yan M, Jiang G, Ge C, Xie H, Wan D, Yang S, Zheng S, Gu J (2007) CD133 positive hepatocellular carcinoma cells possess high capacity for tumorigenicity. Int J Cancer 120(7):1444–1450. https://doi.org/10.1002/ijc.22476

115. Ma S, Chan KW, Hu L, Lee TK, Wo JY, Ng IO, Zheng BJ, Guan XY (2007) Identification and characterization of tumorigenic liver cancer stem/progenitor cells. Gastroenterology 132(7):2542–2556. https://doi.org/10.1053/j.gastro.2007.04.025

116. Yang ZF, Ho DW, Ng MN, Lau CK, Yu WC, Ngai P, Chu PW, Lam CT, Poon RT, Fan ST (2008) Significance of CD90+ cancer stem cells in human liver cancer. Cancer Cell 13(2):153–166. https://doi.org/10.1016/j.ccr.2008.01.013

117. Kim CF, Jackson EL, Woolfenden AE, Lawrence S, Babar I, Vogel S, Crowley D, Bronson RT, Jacks T (2005) Identification of bronchioalveolar stem cells in normal lung and lung cancer. Cell 121(6):823–835. https://doi.org/10.1016/j.cell.2005.03.032

118. Klein WM, Wu BP, Zhao S, Wu H, Klein-Szanto AJ, Tahan SR (2007) Increased expression of stem cell markers in malignant melanoma. Mod Pathol 20(1):102–107. https://doi.org/10.1038/modpathol.3800720

119. Zhang S, Balch C, Chan MW, Lai HC, Matei D, Schilder JM, Yan PS, Huang TH, Nephew KP (2008) Identification and characterization of ovarian cancer-initiating cells from primary human tumors. Cancer Res 68(11):4311–4320. https://doi.org/10.1158/0008-5472.CAN-08-0364

120. Bapat SA, Mali AM, Koppikar CB, Kurrey NK (2005) Stem and progenitor-like cells contribute to the aggressive behavior of human epithelial ovarian cancer. Cancer Res 65(8):3025–3029. https://doi.org/10.1158/0008-5472.CAN-04-3931

121. Szotek PP, Pieretti-Vanmarcke R, Masiakos PT, Dinulescu DM, Connolly D, Foster R, Dombkowski D, Preffer F, Maclaughlin DT, Donahoe PK (2006) Ovarian cancer side population defines cells with stem cell-like characteristics and Mullerian inhibiting substance responsiveness. Proc Natl Acad Sci U S A 103(30):11154–11159. https://doi.org/10.1073/pnas.0603672103

122. Li C, Heidt DG, Dalerba P, Burant CF, Zhang L, Adsay V, Wicha M, Clarke MF, Simeone DM (2007) Identification of pancreatic cancer stem cells. Cancer Res 67(3):1030–1037. https://doi.org/10.1158/0008-5472.CAN-06-2030

123. Xin L, Lawson DA, Witte ON (2005) The Sca-1 cell surface marker enriches for a prostate-regenerating cell subpopulation that can initiate prostate tumorigenesis. Proc Natl Acad Sci U S A 102(19):6942–6947. https://doi.org/10.1073/pnas.0502320102

124. Grange C, Tapparo M, Collino F, Vitillo L, Damasco C, Deregibus MC, Tetta C, Bussolati B, Camussi G (2011) Microvesicles released from human renal cancer stem cells stimulate angiogenesis and formation of lung premetastatic niche. Cancer Res 71(15):5346–5356. https://doi.org/10.1158/0008-5472.CAN-11-0241

125. Bussolati B, Bruno S, Grange C, Ferrando U, Camussi G (2008) Identification of a tumor-initiating stem cell population in human renal carcinomas. FASEB J 22(10):3696–3705. https://doi.org/10.1096/fj.08-102590

126. Driessens G, Beck B, Caauwe A, Simons BD, Blanpain C (2012) Defining the mode of tumour growth by clonal analysis. Nature 488(7412):527–530. https://doi.org/10.1038/nature11344

127. Larsimont JC, Youssef KK, Sanchez-Danes A, Sukumaran V, Defrance M, Delatte B, Liagre M, Baatsen P, Marine JC, Lippens S, Guerin C, Del Marmol V, Vanderwinden JM, Fuks F, Blanpain C (2015) Sox9 controls self-renewal of oncogene targeted cells and links tumor initiation and invasion. Cell Stem Cell 17(1):60–73. https://doi.org/10.1016/j.stem.2015.05.008

128. Tang KH, Dai YD, Tong M, Chan YP, Kwan PS, Fu L, Qin YR, Tsao SW, Lung HL, Lung ML, Tong DK, Law S, Chan KW, Ma S, Guan XY (2013) A CD90(+) tumor-initiating cell population with an aggressive signature and metastatic capacity in esophageal cancer. Cancer Res 73(7):2322–2332. https://doi.org/10.1158/0008-5472.CAN-12-2991

Chapter 5

Informed Consent Issues for Cell Donors

Insoo Hyun

Abstract

Stem cell-based chimera research depends on the free and voluntary provision of human biomaterials necessary for the derivation of pluripotent stem cell lines. Informed consent requirements for the procurement of human embryos, gametes, and somatic cells must take into account unique features of biomedical research involving the use of immortal cell lines that carry their donors' genetic information. The extent and basis for donors' rights, including the right to withdraw from research, are explored here in detail.

Key words Informed consent, Donors' rights, Research withdrawal

1 Introduction

The advancement of chimera research depends on the public's participation in human stem cell research. Public participation consists of the donation or provision of necessary starting biomaterials (i.e., preimplantation embryos, gametes, and somatic cells) for the derivation of stem cell lines used in chimera studies. To ensure that the procurement of these materials is conducted in a manner consistent with current ethical standards for informed consent, and to encourage the implementation of additional stem cell-specific considerations during the consent process, guidelines issued from the International Society for Stem Cell Research (ISSCR) recommend the following [1].

First, consent for the donation of biological materials should be obtained very close to the proposed time in which the materials are to be transferred to the research team. This call for "explicit and contemporaneous consent" requires donors' permission to use their biomaterials to derive immortal stem cell lines. This includes the need to obtain consent from any third-party gamete donors involved in the creation of fertility clinic embryos that may later be used for research. Third-party gamete donors who provide sperm or eggs for assisted reproductive purposes may object to their inadvertent participation in supporting human embryonic stem

Insoo Hyun and Alejandro De Los Angeles (eds.), *Chimera Research: Methods and Protocols*, Methods in Molecular Biology, vol. 2005, https://doi.org/10.1007/978-1-4939-9524-0_5,

cell research, and for this reason they need to be recontacted and consented specifically for their possible complicit involvement in stem cell research.

The only exception to this contemporaneous consent requirement is in cases where researchers obtain somatic cells from a tissue bank to derive new stem cell lines. However, this exception applies only if the tissue bank supplying the somatic cells indicates in its own donor consent forms the possibility that their banked tissues might be used for stem cell research, especially research involving the creation of human embryos (for example, by somatic nuclear transfer or artificial gamete creation, or through a method that might reprogram somatic cells to totipotency).

Other key elements from the ISSCR guidelines include factors that aim to bolster the informed consent dialogue for human material donors. To this end, representatives of the research team must review several important discussion points with potential research donors at the time of the informed consent interview. For example, embryo donors specifically should be told that their donated IVF embryos will be destroyed during the process of human embryonic stem cell derivation. Donors for all types of materials must be informed that the resulting stem cell lines will typically be genetically sequenced and will likely contain all of or partially their own genetic information. Resulting stem cell lines will be shared with researchers at different institutions for other ethically reviewed research purposes, including, importantly, their possible transfer into animal hosts for chimera studies. Donors should also be told whether there are any plans for researchers to share with them commercial royalties resulting from later applications of the derived stem cells (as discussed further below). And all donors must be made aware of their alternatives to donating human materials for research. Potential donors must know that, whatever decision they make, the quality of their own medical care will not be affected.

Although the guidelines from the US National Academy of Sciences (NAS) promulgate a very similar set of informed consent standards [2], the ISSCR guidelines depart from its US counterpart by adding further recommendations for ways regulators and researchers can improve the practice of informed consent. Informed consent should be viewed as an ongoing process of dialogue with potential research donors, not simply a one-time event resulting in the signing of a contractual document. To improve the quality of this interactive dialogue, the ISSCR guidelines recommend that, whenever possible, the person conducting the informed consent interview should have no vested interest in the research protocol. This is to help minimize the possibility of biasing the potential donor's understanding of the risks and benefits of research participation. Also, counseling services should be made available upon request to any human material donor prior to procurement. And all procurement procedures should be reviewed periodically and improved in light of new sociological studies or experiences pertaining to donors' understanding of their participation in stem cell research.

Guidelines such as these are meant to evolve over time as new issues are considered at the periphery of accepted institutional and regulatory practice. One ongoing challenging issue concerns the nature and limits of what it means for human material donors to participate in stem cell research. Does research participation entitle a donor to withdraw later—perhaps very much later—a resulting pluripotent stem cell line from all research uses? Similarly, does research participation mean that an original donor can direct what kind of research can and cannot be performed using a resulting stem cell line? At what stage of the research regulation process should this issue be managed, by scientists during the study design phase of the research, by the institutional review board (IRB) for human subject protections, or by the stem cell research oversight (SCRO) committee? Let us begin by addressing the first question.

Traditionally, research ethics allows research participants the right to withdraw themselves from a scientific project at any time for any reason. But normally this right presumes that human subject research involves ongoing scientific activities performed on the participant's *body* such that research withdrawal is intimately connected to a person's right to bodily integrity and sovereignty (personal autonomy). Typically in these cases, when a human research subject decides to withdraw from a study, the normal course of action is rather straightforward—researchers must stop the experimental intervention on that person and remove him or her from the research project. In most human subject research, such as clinical trials, the issue of subject withdrawal and the conditions necessitating its implementation are usually addressed together by the research team during the design of their study and by the IRB. Although there will always be safety and health monitoring issues and the like to be concerned with, responsible research withdrawal typically does not raise the sort of questions posed by stem cell research regarding what it means *philosophically* for original donors to be involved as "participants" in ongoing tissue culture studies.

Some may point out that similar questions have been asked about what it means for persons to participate in genetic studies involving their private genetic information. Some may argue that participants in genetic studies have a moral right to withdraw their genetic samples from research at any point, even if these samples have been coded to protect their privacy. This position seems to be based on the belief that individuals have an autonomy interest in controlling the uses of their own genetic information. In the same way that an individual can autonomously opt out of research conducted on his or her body at any point, so too can an individual autonomously opt out of allowing researchers to conduct studies on his or her genetic samples. A person's right to autonomy governs, so the argument goes, decisions over the research uses of one's own body as well as the research uses of one's DNA, regardless of whether the DNA samples have been removed long ago from one's physical self.

Although the views above can be controversial, I generally tend to agree with this line of reasoning as it applies to genetic research. However, my chief concern is that there may be key differences between genetic research and pluripotent stem cell research that would make an analogous conclusion harder to draw for the latter.

First, there is the complicating fact that pluripotent stem cell lines may have differing degrees of genetic similarity to their original donors depending on how these lines were derived and whether they were genetically altered for research purposes. The presence of either of these two variables would weaken attempts to extend the genetic study analogy to some types of stem cell research original donors. Why?

Our ability to extend the genetic research analogy to stem cell original donors relies on a simple but important presupposition—namely, that the strength of a donor's rightful claim to withdraw a biological sample from research depends directly on whether that person's genetic information is actually contained in that sample. We assume that, in the case of genetic research, such a corresponding link must exist between the genetic sample and the genetic donor. Thus a donor has a strong right to exclude the use of a genetic research sample only if that sample is a genetic match to him or her. The reason why this (often unstated) dependency relation is important is because the rights claim at issue here is a type of autonomy right. If this general approach to understanding research sample withdrawal is correct, then the following outcomes would result across the range of various pluripotent stem cell derivation techniques.

Embryo donors for stem cell research would have a much reduced claim to autonomously withdraw any embryonic stem cell lines derived from their donated embryos. Unlike genetic donors, whose biological samples are 100% genetically matched to them, no embryo donor can claim to have more than a 50% match to any derived embryonic stem cell line. The situation gets more complicated once we recognize that embryo donation can involve multiple deciding parties, any one of whom may veto the use of an embryo for stem cell research according to current NAS and ISSCR guidelines: (a) the couple for whom the IVF embryo was created and who has dispositional authority over it and (b) any third-party gamete donors whose genetic contribution helped create the embryo. If it is true that the strength of any stem cell withdrawal rights exists in proportion to the degree of personal genetic information contained in the research sample, then this logic would militate against recognizing all stem cell withdrawal rights as being fully comparable to the withdrawal rights of donors in genetic research.

Furthermore, this approach would leave some original donors with no withdrawal rights at all over stem cell lines based on a shared-genetics rationale. Most notably, this limitation would involve IVF treatment couples who use third-party gamete donors for both their sperm and eggs, and somatic cell nuclear transfer (SCNT) egg providers who contribute no personal genetic information to the resulting stem cell line aside from their mitochondrial DNA.

On the other hand, for somatic cell donors in induced pluripotent stem (iPS) cell and SCNT studies, each derived stem cell line would be a genetic match to each somatic cell donor. Thus, the rationale for allowing somatic cell donors to withdraw stem cell lines aligns with the rationale for allowing genetic donors to withdraw their samples from genetic research, assuming that the resulting iPS cell or SCNT stem cell line has not been genetically altered for research purposes.

There are probably other types of arguments aside from the genetic link rationale I outlined above that could support original donors' rights to withdraw stem cell lines. But whatever these other reasons specifically might be, I believe that they all could be reduced to some variant of an individual's right to exercise personal autonomy. If this assertion is correct, then all ethical arguments supporting an original donor's right to withdraw a stem cell line would ultimately have to address the following question.

Does a donor's right to personal autonomy include dispositional rights of withdrawal and control over biological samples that have been radically transformed in the process of pluripotent stem cell derivation and cultivation? In order for any autonomy-based argument to support either research withdrawal or donor control over future uses, there must exist a strong metaphysical connection between the individual and the research sample in question. In the case of genetic research, the connection between these two is rather tight. But in the case of pluripotent stem cell research (much of which involves quite radical biological transformations) the exact nature of the metaphysical connection between donor and research sample gets thrown into doubt. One of the chief scientific benefits of tissue culture research is that scientists can take a one-time biological donation and serially culture the sample to produce several future generations of research material without ever having to go back to the original donor. In this way, tissues in culture begin to take on a life of their own quite apart from the original donor. In the case of pluripotent stem cell research, cells in culture can be manipulated very far to a point where they are no longer the same type of cells or tissue as the original sample. For example, an iPS cell line can undergo more than 45 doublings of cells, and most of these clonal iPS cells can be differentiated into a wide array of cell types. Does the somatic cell donor have reach-through rights over all these resulting cells?

Some people may argue that having the same nuclear genetic profile to a stem cell line is sufficient to justify a somatic cell donor's reach-through rights to withdraw or control an SCNT or iPS cell line. But this reasoning may be too hasty. What about the ongoing laboratory interventions and developmental pathway manipulations that are necessary for pluripotent stem cells to be maintained in culture? Without these constant interventions and manipulations, pluripotent stem cells would cease to exist physically as an object of study. Should these necessary and coordinated efforts carry any weight in one's determination of whether somatic cell donors have an exclusive right to withdraw stem cell lines? Perhaps they do, for the presence of these dynamic influences makes stem cell research disanalogous to plain genetic research *per se*. Some philosophers may go so far as to defend stem cell researchers by invoking a vaguely Lockean argument according to which researchers have accrued considerable private property interests in their derived stem cell lines because they have already "mixed their labor" with these cultured cells to a very high degree.

This is a complicated issue. Both somatic cell donors and stem cell researchers might each reasonably claim to have property rights over the derived stem cell lines. Deciding who has the stronger claim will depend on which factor one judges to be determinative, having a genetic link to the stem cells or having established a record of creative technical labor to bring these cells into being. This issue is further complicated by the likelihood that clinical translational researchers may need to maintain long-term donor tracking for health information that might be relevant for the safe clinical use of any resulting stem cell lines and their derivatives. If somatic cell donors are to be burdened with this request as a condition of their participation in stem cell research, then it would seem difficult to justify not granting these donors some degree of control, over future research use. The trick is to know how to balance all of these considerations into a single coherent and justifiable policy.

Some observers may point out that all of this could be sorted out through the informed consent process. There is some truth to this. For example, one approach could be to encourage researchers to disclose their intention of only enrolling donors who agree to waive their right to withdraw any derived stem cell lines from research use, as well as their right to control their future research uses (assuming that such rights exist). The wording of the consent form could be expressed as follows: "Although some research teams may allow original donors to withdraw derived stem cell lines from research use at any point in the future, or to control their future research uses, our research team acknowledges that we cannot guarantee either of these possibilities for our donors. If you are uncomfortable with our position on this matter, you should do not agree to participate as a human materials donor for our project."

Wording such as this should not be interpreted as a sneaky attempt by researchers to shrug off important moral rights of the donor. In many cases, stem cell researchers cannot control how researchers at other institutions will ultimately use the derived stem cells or their direct derivatives; neither can they force other researchers to stop using the derived cells, especially if other researchers have dramatically transformed these cells using methods they have patented in their own laboratories. These realities present significant practical limitations on what a research team can promise to an original donor during the consent process. It may be misleading to suggest to a potential donor that he or she can actually retain autonomous control over a resulting stem cell line once it has been shared with others. Making such a promise to donors may, at best, be an empty gesture. At worst, obtaining informed consent from a donor on the basis of such a shaky promise may be morally pernicious, for it could undermine the authenticity of the informed consent process.

Some critics may argue that these practical limitations can be overcome. Some might claim that a research team can easily provide room for their original donors' preferences in the material transfer agreements (MTAs) that accompany each stem cell line distributed to other researchers. However, those on the other side of the issue may object that it would be too burdensome to try to monitor and enforce every single one of these donor preferences stipulated in the language of the MTA. Others may raise the more philosophical argument that allowing donor restrictions on the specific uses of derived stem cell lines runs counter to the spirit of "open science" and open research use that is sanctioned within the ISSCR guidelines. The ability for scientists to pursue ethically reviewed research that they, not the original donors, deem to be important is crucial for the advancement of socially beneficial scientific research. This last brand of argument essentially pits the value of broad social benefit (beneficence) against the autonomy rights of original donors to control stem cell lines (assuming that such autonomy rights exist). Whether or not the practical limitations of honoring donor preferences can be resolved through carefully worded material transfer agreements does nothing to address this deeper philosophical controversy.

There are limitations to what informed consent documents and MTAs can resolve through their structure and wording. The careful wording of these documents does not itself disarm underlying philosophical questions about whether the conditions spelled out therein are ethically justified. It is possible for people to sign unfair informed consent forms and MTAs. Signing these documents does not automatically render unfair contracts fair. In order to know how to separate fair contracts from unfair contracts, we first have to wrestle with underlying ethical questions concerning the range of an original donor's autonomy rights over radically manipulated stem cell lines and their derivatives, and the ethical justifications supporting researchers' unrestricted use of these lines. The practical wording of documents cannot resolve these philosophical questions, even if ethical review committees allow these documents to be used for stem cell research.

Despite these novel philosophical and practical challenges, stem cell-specific informed consent standards, as promulgated most notably in the ISSCR guidelines, offer researchers, regulators, and institutions reasonably clear guidance for how to best ensure that the procurement of human biomaterials is conducted responsibly for the derivation of stem cell lines used in chimera research and other downstream stem cell-related studies [3].

References

1. International Society for Stem Cell Research (2016) Guidelines for stem cell science and clinical translation. http://www.isscr.org/docs/default-source/all-isscr-guidelines/guidelines-2016/isscr-guidelines-for-stem-cell-research-and-clinical-translationd67119731dff6ddbb37cff0000940c19.pdf?sfvrsn=4. Accessed 20 Oct 2018
2. National Academy of Sciences (2005) Guidelines for human embryonic stem cell research. The National Academies Press, Washington, DC
3. Daley GQ, Hyun I, Apperley JF et al (2016) Setting global standards for stem cell research and clinical translation: the 2016 ISSCR guidelines. Stem Cell Reports 6:787–797

Part II

Non-Human Hosts: Species and Developmental Stages

Chapter 6

Chick Models and Human-Chick Organizer Grafts

Iain Martyn, Tatiane Y. Kanno, and Ali H. Brivanlou

Abstract

The combination of affordability, large size, and ease of access at almost every stage of development renders the chick an excellent model organism for studying vertebrate development. Not only is it a great system in and of itself, but these qualities make it a great host for interspecies chimera experiments. In this chapter we highlight some notable examples of mammalian-chick chimeras, and show how one can for instance use the chick to push mammalian stem cell experiments further to learn about the behavior and capabilities of these cells in vivo. In particular, here we present the methodology necessary for transplantation of human embryonic stem cell (hESC)-derived "gastruloids" stimulated to generate a human organizer into the chick embryo. In these human-chick chimeras, the human organizer cells self-organize to contribute directly to notochord-like tissue and indirectly induce host chick cells to generate neural tissue.

Key words Chicken embryos, Human organizer, Human embryonic stem cells (hESC), Chimeras, Transplantation

1 Introduction

The great advantages of the chick model for studying developmental biology are its affordability, versatility, and ease of access [1, 2]. Armed with only an incubator and a nearby farmyard or poultry facility one can examine almost any stage of vertebrate development, from pre-gastrulation to neurulation to hatching, simply by cracking open an egg at the appropriate time and examining its contents. The chick has precise classifications of embryonic stage [3], and is amenable to classic experimental embryological manipulations, such as tissue grafting, ablation, tissue recombination, and genetic perturbations. Due to its transparency the chick embryo can also be imaged live in ovo or ex ovo using simple culture systems. Chick embryos are excellent models for interspecies transplantation studies, providing a technically less challenging platform than the mammalian counterpart.

Indeed, the chick experimental model has been used for decades as a xenograft host for chimera experiments [4–8], beginning with the classical work of Nicole le Douarin and colleagues who

Insoo Hyun and Alejandro De Los Angeles (eds.), *Chimera Research: Methods and Protocols*, Methods in Molecular Biology, vol. 2005, https://doi.org/10.1007/978-1-4939-9524-0_6,

grafted quail neural crest cells at the neural plate boundary of the chick embryo, generating chick-quail chimeras [9–11]. Successful transplantations of mammalian cells and tissues into the chick embryos have also been reported and are facilitated by the fact that they can be grown at the same temperature. Mouse-chick chimeras have been used to understand and dissect the mechanism of specific cell types and tissue such as neural crest, motor neurons, and somite specification [12–14]. Mammalian cells thus respond to local signaling and contribute to the chick anatomy. Finally, human embryonic and adult stem cells have been previously transplanted in chick embryos and shown to be adaptable to local signals [5, 15, 16].

We have recently shown that the chick system can also be used to study very early human development, which remains a final frontier in developmental biology. Our little current understanding of our own development mostly comes from the Carnegie and Kyoto Embryo Collection [17, 18] and is primarily based on morphology and a handful of molecular markers. This is due to the paucity of source of biological material and ethical limitations. To overcome these limitations human embryonic stem cells (hESCs) have been used to generate artificial human embryos, also known as gastruloids [19, 20], which can be generated in vast quantities. As powerful a platform as gastruloids are, they are still self-organizing cells, grown in in vitro culture, and thus at best conclusions derived from the experiments will only unveil what hESCs can do, not necessarily what they would do in an embryonic ecosystem in vivo. It is thus crucial to assess to what extent the in vitro results can be validated in vivo. Interspecies chimeric assays have been used in the past to validate hESC results from cell culture, including our own work which featured the first human-mouse chimera [21].

Here we present the methodology necessary for transplantation of hESC-derived gastruloids stimulated to generate a human "organizer" into chick embryo. In these human-chick chimeras organizer cells self-organize to contribute directly to notochord-like tissue and indirectly induce host cells to generate neural tissue [22].

2 Induction of Secondary Ectopic Axis by Transplantation of hESC-Derived Human Organizer

This method was originally developed with mouse embryonic fibroblast conditioned medium (MEF-CM) using our own hESC line, RUES2, and commercially available CYTOO micropatterned chips. The reader may also find it useful to refer to Chapman et al. [23] for further detail on the preparation of the chick host, to Stern [24] for further detail on the mechanics of donor transfer and placement within the host, and to Deglincerti et al. [20] for further detail on hESC micropattern techniques, as our method was initially based on these sources.

2.1 Materials for hESC Micropattern Culture

1. MEF-CM.
2. ROCK-Inhibitor (Y-27632 dihydrochloride) (Abcam, cat. no. ab120129).
3. Penicillin–Streptomycin (Life Technologies, cat. no. 15140-148).
4. Accutase (StemCell Technologies, cat. no. 07920).
5. Laminin-521 (BioLamina, cat. no. LN521-04).
6. PBS++.
7. PBS−−.
8. WNT3A stock solution:

   Mix 987 μL of PBS−− with 13 μL of bovine albumin fraction V 7.5% solution (Thermo Fisher Scientific, cat. no. 15260037). Use 500 μL of this to dissolve 10 μg of recombinant mouse WNT3A protein (R&D Systems, cat. no. 1324-WN-010) to a final concentration of 20 μg/mL. Prepare 20 μL aliquots in microcentrifuge tubes and store them at −80 °C for up to 6 months. Thawed aliquots can be stored at 4 °C for 2 weeks.
9. Activin stock solution:

   Resuspend 10 μg of recombinant activin A protein (R&D, cat. no. 338-AC-050/CF) with 100 μL of sterile 4 mM HCl. Prepare 10 μL aliquots in microcentrifuge tubes and store them at −80 °C for up to 6 months. Thawed aliquots can be stored at 4 °C for 2 weeks.
10. bFGF solution:

    Resuspend bFGF (Life Technologies, cat. no. PHG0263) in PBS−− containing 0.1% (wt/vol) BSA to a final concentration of 20 μg/mL. Prepare 100 μL aliquots in microcentrifuge tubes and store them at −80 °C for up to 6 months. Thawed aliquots can be stored at 4 °C for 2 weeks.
11. 35 mm tissue culture dishes.
12. Parafilm "M."
13. Hemocytometer.
14. CYTOO chips, Arena 500 A (CYTOO, cat. no. 10-024-00-18).
15. Coverslip forceps (Fine Science Tools, cat. no. 11251-33).
16. 15 and 50 mL conical centrifuge tubes.

17. Pipette controller (Accujet Pro, BrandTech, cat. no. 26333) and 5, 10, and 25 ml serological pipettes.
18. 2, 10, 20, 200, and 1000 μL pipettes and corresponding sterile pipette tips.
19. 10 μL barrier pipette tips.
20. CO_2 incubator with controlling and monitoring system for CO_2, humidity, and temperature (HeraCell 150i; Thermo Fisher Scientific, cat. no. 51026282).
21. Biosafety cabinet for cell culture (SterilGARD III Advance SG403, The Baker Company).
22. Cell culture centrifuge.

2.2 Materials for Grafting

1. Dissecting microscope with transmitted light base.
2. Pannett-Compton saline:

   Solution A: 121 g NaCl, 15.5 g KCl, 10.42 g $CaCl_2 \cdot 2H_2O$, 12.7 g $MgCl_2 \cdot 6H_2O$, H_2O to 1 L.

   Solution B: 2.365 g $Na_2HPO_4 \cdot 2H_2O$, 0.188 g $NaH_2PO_4 \cdot 2H_2O$, H_2O to 1 L; before use, mix in order 120 mL Solution A, 2700 ml H_2O, and 180 mL Solution B.
3. Saline solution (7.19 g NaCL/1 L distilled water).
4. Bacto-Agar.
5. Kimwipes.
6. Embryo forceps (Fine Science Tools, cat. no. 11252-20).
7. Fine curved scissors (Fine Science Tools, cat. no. 14061-09).
8. 35 mm μ-Dish, high (optical plastic) Ibidi dishes: These dishes enable the fluorescently tagged donor hESCs to be imaged live with good resolution. If this is not necessary regular 35 mm tissue culture dishes will suffice.
9. 150 × 25 mm tissue culture dish.
10. Filter paper frames made by cutting Whatman Filter Paper #2 into 35 mm diameter circles and then either hole-punching clover leaf shape out of the center or cutting out an inner circle.
11. Container for egg waste.
12. Flexible transfer pipets (Fisherbrand 13-711-7M).
13. Tungsten needles 0.25 mm diameter (Fine Science Tools, cat. no. 10130-10): We typically fix these to a BD PrecisonGlide Needle using superglue and then mount the needle onto a BD 1 mL syringe for manipulation.
14. 38 °C incubator.
15. Fertilized White Leghorn eggs incubated 12–18 h (depending on the stage of development required).

2.3 Materials for Fixing, Staining, and Imaging

1. Normal donkey serum (Jackson ImmunoResearch, cat. no. 017-000-121).
2. Tween-20.
3. Primary antibodies:

   Anti-SOX2, R&D AF2018, 1:200 dilution, works in chick and human; anti-BRACHYURY, R&D Systems AF-2085, 1:300 dilution, works in chick and human; anti-human nuclear antigen, Novus Biologicals NBP2-34525AF647, works in human only.
4. Secondary antibodies:

   Corresponding secondary donkey Alexa Fluorophore-conjugated antibodies. We recommend using longer wavelength fluorophores (i.e., 555, 594, and 647) as there is often significant autofluorescence around 488 nm.
5. Paraformaldehyde 4% (wt/vol) in PBS, pH 7.4.
6. Triton X-100 detergent.
7. Fluoromount-G mounting medium (Southern Biotech, cat. no. 0100-01).
8. Blocking solution:

   Add 10 μL of Triton X-100 and 300 μL of normal donkey serum to 10 mL of PBS. Gently mix by inversion. The solution can be stored at 4 °C for 1 week.
9. Washing solution:

   Add 20 μl of Tween-20 to 20 mL of PBS. Gently mix by inversion. The solution can be stored at room temperature (18–25 °C) for 6 months.
10. DAPI stock:

    Reconstitute 1 mg of DAPI (Cell Signaling Technologies, cat. no. 4083S) in 10 ml of deionized water to obtain a 0.1 mg/mL solution. Make 100 μL aliquots and freeze them at −20 °C for up to 2 years.
11. Inverted fluorescence microscope (Olympus IX83) with an XY motorized stage for tiling and digital imaging capture system (Andor Zyla 4.2 C-Mos camera).
12. Inverted laser scanning confocal microscope (Zeiss LSM780) for high-resolution optical slicing.

3 Method

3.1 Graft Preparation

1. Cover the base of a 150 × 25 mm tissue culture dish with Parafilm, taking care to put the more sterile side (side originally facing the packaging paper) facing up.

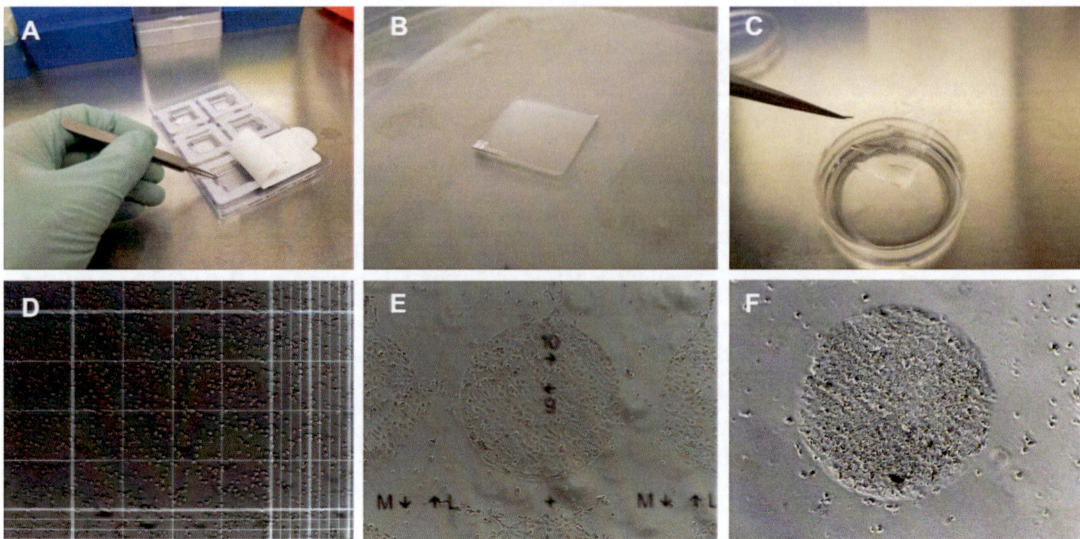

Fig. 1 (**a**, **b**). Coating CYTOO chip with laminin. (**c**) After coating the micropatterns are visible on the CYTOO chip by eye. (**d**) Counting cells with a hemocytometer. (**e**) 500 μm diameter micropattern 1 h after seeding with hESCs. (**f**) 500 μm diameter micropattern 18 h after seeding of hESCs, ready for WNT3A+ACTIVIN treatment

2. Dilute Laminin-521 into PBS++ to a final concentration of 10 μg/mL. Pipette at least 400 μL of this solution onto the Parafilm, letting the solution bead into a single drop.
3. Using the coverslip forceps place the CYTOO chip pattern side down on the droplet (Fig. 1a, b). The side of the CYTOO chip on which the CYTOO label is written in the forward direction is the patterned surface. If necessary gently press the borders of the chip down with the tweezers to ensure that the entire surface of the chip is wetted.
4. Incubate the chip at 37 °C for 3 h or overnight at 4 °C.
5. Wash the chip three times with PBS++. Care should be taken to avoid drying the surface of the chip between wash cycles. After the last wash place CYTOO chip pattern side up in a new 35 mm tissue culture dish filled with 2 mL of PBS++ (Fig. 1c).
6. Take a 35 mm tissue culture dish that is 60–80% confluent with hESCs, remove media, and add 1 ml of Accutase. Note that each CYTOO chip requires 1×10^6 cells for seeding, and a 60–80% confluent 35 mm dish should contain $1–2 \times 10^6$ cells.
7. Return the dish to the incubator for 7 min while the Accutase takes effect.
8. Take dish from incubator and gently swirl the dish, checking to make sure that all cells are dissociated. Pipet the dissociated cells into a 15 mL conical centrifuge tubes, along with 1 mL of prewarmed MEF-CM media.

9. Centrifuge the suspension at 300 × *g* for 4 min at room temperature and discard the supernatant. Resuspend the pellet in 5 mL of PBS−− and then spin down again.
10. Discard the supernatant and resuspend in 1 mL of MEF-CM + ROCK-Inhibitor + Pen-Strep. Aggregates formed during centrifugation must be broken up.
11. Count cells using a hemocytometer (Fig. 1d).
12. Pipet 1 × 10^6 cells into a new conical centrifuge tube and adjust the volume to 2 mL with MEF-CM + ROCK-Inhibitor + Pen-Strep media.
13. Aspirate PBS++ from coated CYTOO chip and immediately add the total 2 mL volume of hESCs.
14. Place CYTOO chip in the cell culture incubator, taking care not to swirl the dish or otherwise concentrate cells in the center.
15. After 1 h replace the media with MEF-CM + Pen-Strep (Fig. 1e). Incubate overnight for 12–16 h (*See* **Note 1**).
16. Transfer the seeded chip to a new 35 mm tissue culture dish and add 2 mL of MEF-CM + Pen-Strep with 100 ng/mL WNT3A and 100 ng/mL activin A to start differentiation (Fig. 1f).
17. Incubate in cell culture incubator for 24–30 h.

3.2 Host Preparation

1. Incubate fresh fertilized White Leghorn eggs at 37 °C and 50% humidity for 10–16 h to reach the appropriate stage (HH 2-3) (*See* **Note 2**).
2. Make agar mounts by first microwaving 70 mL of saline solution with 0.6 g of Bacto-Agar and then mixing 1:1 with fresh thin albumin. Care should be taken so that all Bacto-Agar is dissolved in microwave step. Keep the agar solution and albumin mix at 40 °C. Add Pen-Strep, and then use the serological pipettes to add 1 mL to each dish in such a way that the solution does not contact the edge and stays beaded up in the center to form a mount.
3. Remove eggs from incubator and rest at room temperature for at least 30 min.
4. Crack the egg into a high-walled 150 × 25 mm tissue culture dish, taking care when cracking that the embryo side is up. Note that during incubation the embryo will always be facing upwards due to the chalazae, a strand of protein which holds the egg yolk in place.
5. Using Kimwipes or other absorbent paper gently remove the thick albumin from the area around the embryo (Fig. 2a). Do not attempt to remove the albumin from the region directly on top of the embryo.

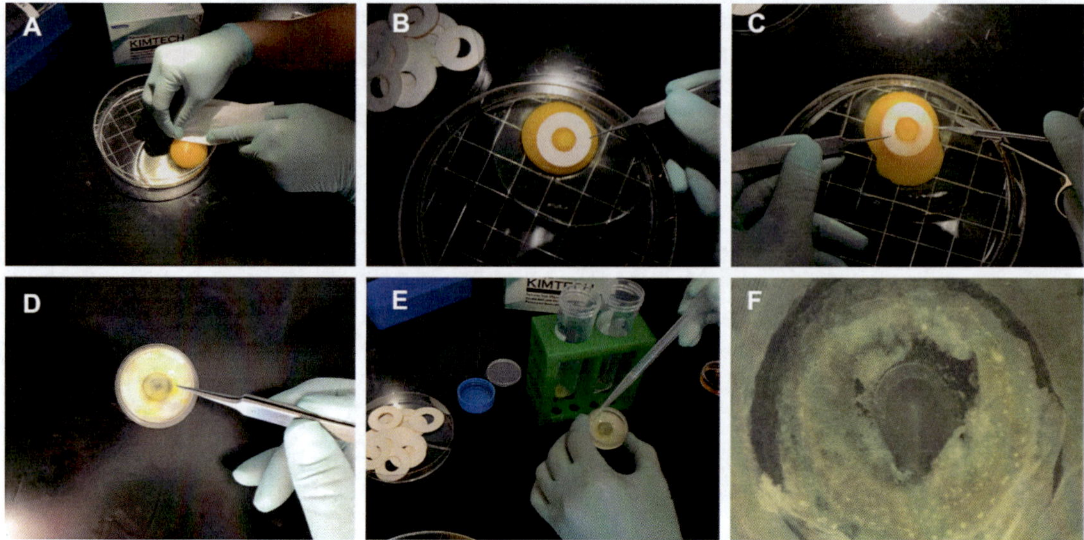

Fig. 2 Preparation of host embryo. (**a**) Cleaning albumin from area around the embryo. (**b–d**) Attaching filter paper frame, cutting vitelline membrane around the frame, and removing frame from yolk and placing in 35 mm dish with agar mount. (**e**) Washing the EC culture host. (**f**) Ventral view of HH 3 stage EC culture embryo ready for donor graft

6. Using embryo tweezers place the prepared filter paper frame over the egg so that the embryo is positioned in the central window region. Let the filter paper adhere to the membrane (Fig. 2b).
7. Using fine scissors cut around the outside of the frame. The embryo will be stretched out within the frame and remain so throughout processing.
8. Using tweezers, gently lift the filter paper frame off the embryo at a 45° angle from the point where the yolk first began to flow upon cutting (Fig. 2c).
9. Place the embryo with the ventral side facing up in one of the prepared 35 mm Ibidi or tissue culture dishes (Fig. 2d). The embryo should be directly over the top of the central agar-albumin mount so that residual liquid flows toward the sides of the dish.
10. Using a flexible Pasteur pipette, gently wash the embryo and filter paper frame with Pannett-Compton solution to remove as much yolk as possible (Fig. 2e). Make sure during the washing step that the embryo remains attached to the vitelline membrane and is not injured.
11. Remove all excess liquid and place in incubator; the embryo is ready to receive the graft.

3.3 Grafting and Induction

1. Fill three 35 mm tissue culture dishes and a 50 mL Falcon tube with Pannett-Compton + Pen-Strep solution. Place in the sterile derivation hood ready for use.
2. Remove hESC WNT3A+ACTIVIN stimulated micropatterns and EC chick embryos from their respective incubators and bring to the sterile derivation hood.
3. Place the EC culture chick embryo under the dissecting microscope and select a potential graft site in the marginal zone between the area pellucida and area opaca approximately 90° away from the site of primitive streak initiation. If necessary, gently wash the embryo again with Pannett-Compton + Pen-Strep solution and remove yolk.
4. Having identified a suitable graft site, use a tungsten needle to gently score and lift up a portion of the flap of yolky cells (germ wall margin) that covers the inner margin of the area opaca. Work outwards from the area pellucida and avoid penetrating the ectoderm underneath. This will produce a pocket into which the graft can be inserted. As an alternative, just carefully score the endoderm at this margin to create a micropattern size hole.
5. Now set the host aside for the moment and bring the hESC WNT3A+ACTIVIN stimulated micropatterns under the microscope. Identify a suitable micropattern, and, holding the chip down with tweezers in one hand, use a separate tungsten needle to carefully peel off the colony (Fig. 3a, b). The colony should come off the coverslip in one piece with minimum

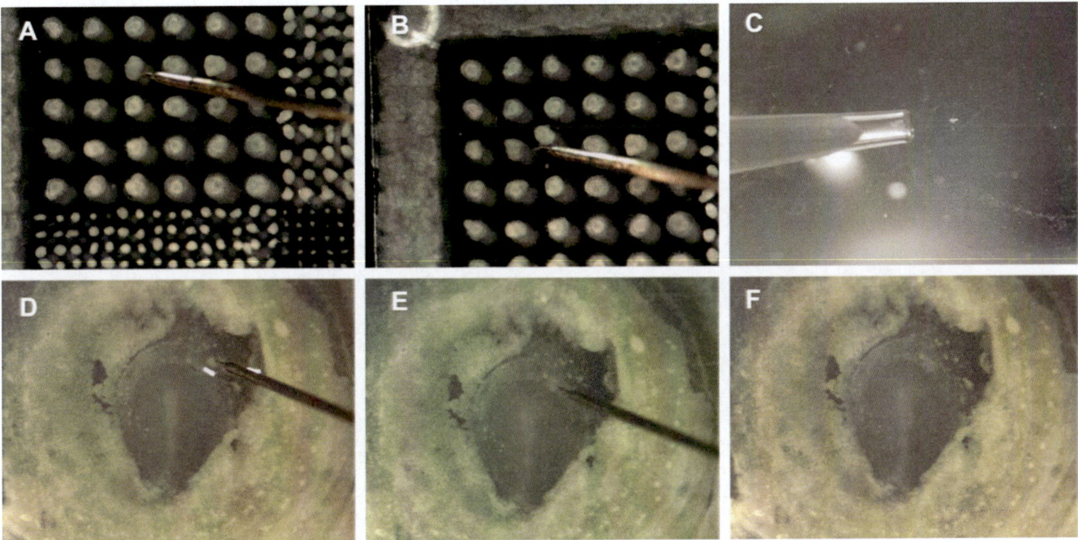

Fig. 3 Grafting manipulations. (**a**) Gently work around the edges of a suitable micropattern to detach it from the coverslip. (**b**) Detached free floating micropattern. (**c**) Positioning the washed detached micropattern at the bottom of a 10 μL pipette tip. (**d–f**) Locate the transferred micropattern and slide it into the pocket

injury. If it is difficult to remove it in one piece or it rolls up upon detachment this is a sign that the colony is not dense enough (i.e., starting density at seeding was too low or not enough cells initially attached properly). Each colony should cohere nicely and preserve its structure upon detachment.

6. Using a 10 μL pipette tip, transfer the donor colony sequentially to a new 35 mL dish filled with Pannett-Compton + Pen-Strep. Repeat this twice more to thoroughly wash the micropattern and make sure that one is not transporting culture media to the chick (Fig. 3c).
7. Once washed, suck up the colony in a new 10 μL pipette tip. Do so by first sucking up 9 μL of the Pannett-Compton solution and then in the last 1 μL suck up the colony. This will position the colony as close to the end of the pipette tip as possible and allow easier placement of the donor in the host site.
8. Remove the micropattern dish from the microscope and replace with the prepared host dish. Identify a clear, unobstructed region of the area pellucida and bring the pipette tip directly over top as close to the embryo as possible without touching it. Gently expel the pipette tip until the donor micropattern is released and successfully transferred (Fig. 3d). This step takes practice since the micropattern is almost invisible against the host epiblast and so is easy to lose. Not preparing the colony properly in the pipette tip or pushing out too quickly or onto a region with too many other features leads to the donor being lost. If the donor cells are fluorescently tagged, it may be possible to locate them under a fluorescent microscope.
9. Using the tungsten needle used to create the pocket, carefully manipulate the donor to the graft site and slip it into the pocket (Fig. 3e, f).
10. Carefully remove any remaining solution. It is important that the embryo and the inside of the ring remain completely dry during incubation. Observe the graft to make sure that it does not become dislodged.
11. If donor cells are fluorescently tagged or otherwise marked, imaging can be performed by placing the culture dish on an inverted microscope to generate tiling that includes the host primitive streak and grafted hESC colony.
12. Once imaged, securely fasten the Ibidi lid and place the dish onto a new 150 × 25 mm tissue culture dish.
13. Once all Ibidi dishes are loaded, wrap the 150 × 25 dish securely with Parafilm and place in the chick incubator for 24 h.

3.4 Analysis of Results

To assess the in vivo properties of the human organizer, whole-mount immunofluorescence and in situ hybridization are performed with antibodies and probes for early neural markers.

HNA is a specific human antibody used to detect the human cells, and SOX2 and SOX3 (in situ probes, kindly provided by F.M. Vieceli) are early neural progenitor markers. Analysis of these molecular marker expression through confocal cross sectioning established that the human cells can induce neural tissue in the chick, since SOX2 and SOX3 are ectopically expressed in chick cells surrounding the grafted human organizer [22].

Methods for immunofluorescence and in situ hybridization can be found in more detail in various publications [25–29]. See Notes as well for additional details. Using the described approach, we observe that the human cells can mix and mingle with the chick cells giving rise to the chick-human chimera. The rate of survival of the grafted cells is around 90%, and of those approximately 60% present an ectopic axis expressing SOX2 and or SOX3.

In conclusion, our in vivo platform is a useful tool to validate results obtained in an in vitro gastruloid system and may be generally applicable to test and explore other aspects of early human development.

4 Notes

1. Problems with hESC attachment to CYTOO chips such as cells attaching outside of the micropatterns or inefficiently attaching within the mircopatterns can sometimes occur. These are most often due to batch-to-batch variation, and can be fixed by changing the working dilution of laminin (test in the range of 5–20 μg/mL), changing the coating time (1–3 h), or changing the batch.
2. Fertilized eggs can be stored at room temperature and are usually viable for 1 week after laid.
3. It is important to keep the embryos as flat as possible during fixation step. In case the embryo detaches from the filter paper, remove all PBS and flatten it on the bottom of the dish, then drop 4% paraformaldehyde in PBS on top of the embryo. Alternatively, embryos can be pinned out on a Sylgard dish prior to fixation.
4. Trim the tissue surrounding the embryo before proceeding to in situ hybridization or immunofluorescence and subsequent manipulations.
5. For in situ hybridization, solutions from fixing to hybridization should be RNase free (except PFA).

6. Optimize the time of proteinase K incubation. The incubation time will depend on the embryo stage and the enzymatic activity of proteinase K. It should be determined empirically to achieve the best signal and maintain the embryo morphology. Embryos until stage HH3 do not need to be incubated with proteinase K. Embryos are very fragile after proteinase K treatment, so make sure to change solutions very carefully.

References

1. Stern CD (2005) The chick: a great model system becomes even greater. Dev Cell 8 (1):9–17
2. Mok GF, Alrefaei AF, McColl J, Grocott T, Münsterberg A (2015) Chicken as a developmental model. eLS [Internet] (January). http://doi.wiley.com/10.1002/9780470015902.a0021543. p. 1–8
3. Hamburger V, Hamilton HL (1951) A series of normal stages in the development of the chick embryo. J Morphol 88(1):49–92
4. Goldstein RS (2010) Transplantation of human embryonic stem cells and derivatives to the chick embryo. In: Turksen K (ed) Human embryonic stem cell protocols. Humana Press, Totowa, NJ, pp 367–385
5. Lee G, Kim H, Elkabetz Y, Al Shamy G, Panagiotakos G, Barberi T et al (2007) Isolation and directed differentiation of neural crest stem cells derived from human embryonic stem cells. Nat Biotechnol 25(12):1468–1475
6. Boulland J-L, Halasi G, Kasumacic N, Glover JC (2010) Xenotransplantation of human stem cells into the chicken embryo. J Vis Exp [Internet] 41:11–15
7. Glover JC, Boulland J-L, Halasi G, Kasumacic N (2010) Chimeric animal models in human stem cell biology. ILAR J [Internet] 51 (1):62–73
8. Li W, Huang L, Lin W, Ke Q, Chen R, Lai X et al (2015) Engraftable neural crest stem cells derived from cynomolgus monkey embryonic stem cells. Biomaterials 39:75–84
9. Le Douarin NM, Jotereau FV (1975) Tracing of cells of the avian thymus through embryonic life in interspecific chimeras. J Exp Med 142 (1):17–40
10. Couly GF, Le Douarin NM (1985) Mapping of the early neural primordium in quail-chick chimeras. I. Developmental relationships between placodes, facial ectoderm, and prosencephalon. Dev Biol 110(2):422–439
11. Le Douarin N, Dieterlen-Lièvre F, Creuzet S, Teillet MA (2008) Chapter 2: Quail-chick transplantations. Methods Cell Biol 87 (08):19–58
12. Fontaine-Pérus J, Halgand P, Chéraud Y, Rouaud T, Velasco ME, Cifuentes Diaz C et al (1997) Mouse-chick chimera: a developmental model of murine neurogenic cells. Development 124(16):3025–3036
13. Fontaine-Pérus J (2000) Mouse-chick chimera: an experimental system for study of somite development. Curr Top Dev Biol 48:269–300
14. Fontaine-Pérus J, Chéraud Y (2005) Mouse-chick neural chimeras. Int J Dev Biol 49 (2–3):349–353
15. Jiang X, Gwye Y, McKeown SJ, Bronner-Fraser M, Lutzko C, Lawlor ER (2009) Isolation and characterization of neural crest stem cells derived from in vitro–differentiated human embryonic stem cells. Stem Cells Dev 18(7):1059–1071
16. Lee H, Shamy GA, Elkabetz Y, Schofield CM, Harrsion NL, Panagiotakos G et al (2007) Directed differentiation and transplantation of human embryonic stem cell-derived motoneurons. Stem Cells 25(8):1931–1939
17. O'Rahilly R, Muller F (1987) Developmental stages in human embryos: including a revision of Streeter's "Horizons" and a survey of the Carnegie collection. Carnegie Institution of Washington, Washington, DC
18. Hill M, Shiota K, Yamada S, Lo C (2016) Kyoto embryo collection. iBooks, UNSW, Australia
19. Warmflash A, Sorre B, Etoc F, Siggia ED, Brivanlou AH (2014) A method to recapitulate early embryonic spatial patterning in human embryonic stem cells. Nat Methods 11 (8):847–854
20. Deglincerti A, Etoc F, Guerra MC, Martyn I, Metzger J, Ruzo A et al (2016) Self-organization of human embryonic stem cells on micropatterns. Nat Protoc 11 (11):2223–2232
21. James D, Noggle SA, Swigut T, Brivanlou AH (2006) Contribution of human embryonic

stem cells to mouse blastocysts. Dev Biol 295 (1):90–102

22. Martyn I, Kanno TY, Ruzo A, Siggia ED, Brivanlou AH (2018) Self-organization of a human organizer by combined Wnt and Nodal signalling. Nature 558(7708):132–135
23. Chapman SC, Collignon J, Schoenwolf GC, Lumsden A (2001) Improved method for chick whole-embryo culture using a filter paper carrier. Dev Dyn 220(3):284–289
24. Stern CD (1999) Grafting Hensen's node. Methods Mol Biol 97:245–253
25. Streit A, Stern CD (2001) Combined whole-mount in situ hybridization and immunohistochemistry in avian embryos. Methods 23 (4):339–344
26. Acloque H, Wilkinson DG, Nieto MA (2008) Chapter 9: In situ hybridization analysis of chick embryos in whole-mount and tissue sections. Methods Cell Biol 87(08):169–185
27. Psychoyos D, Finnell R (2008) Double whole mount in situ hybridization of early chick embryos. J Vis Exp 20(20):8–10
28. Psychoyos D, Finnell R (2009) Method for whole mount antibody staining in chick. J Vis Exp 24:5–7
29. Vieceli FM, Simões-Costa M, Turri JA, Kanno T, Bronner M, Yan CYI (2013) The transcription factor chicken Scratch2 is expressed in a subset of early postmitotic neural progenitors. Gene Expr Patterns 13 (5–6):189–196

Chapter 7

The Engraftment of Lentiviral Vector-Transduced Human CD34+ Cells into Humanized Mice

Yoon-Sang Kim, Matthew Wielgosz, and Byoung Ryu

Abstract

Humanized mouse models have been developed to study human hematopoiesis and therapeutic application of hematopoietic stem cell transplantation. To evaluate the safety and efficacy of lentiviral vectors for gene therapy, human CD34+ cells have been transduced with lentiviral vectors and transplanted into the humanized mice. Recipient mice are monitored over time and sacrificed for bone marrow analyses with regard to human cell engraftment, lineage distribution, and vector transduction. This chapter details the procedure for lentiviral transduction and transplantation of human hematopoietic stem/progenitor cells into humanized mice to study inherited human hematological disorders.

Key words Hematopoietic stem cells (HSC), Transplantation, Lentiviral vector, Transduction, Engraftment, Humanized mouse models

1 Introduction

Several mouse strains have been developed to engraft human cells without rejection. These so-called humanized mice are valuable tools to study human cancer, infectious diseases, and hematopoiesis [1, 2]. In particular, xenotransplantation is a useful model system for the development of hematopoietic stem cell (HSC) gene therapy as the recipient mice can support various human hematopoietic lineages that are functional to some extent. Since the first humanized mouse, few genetic modifications have been introduced that overcome limitations of earlier humanized versions. For example, NOD-*scid* mice develop thymic lymphoma [1, 3], which is the major cause of death for this strain. By introducing a complete null mutation (knockout) of interleukin 2 receptor gamma chain (*Il2rg*) gene, the new NSG mouse strain does not develop lymphoma and live longer than their parental NOD-*scid* strain. In addition, NSG mice completely lack immune cells (T, B, and NK cells), and are better hosts for human hematopoietic stem cells than the parent NOD-*scid* strain. NSG mice have been widely used in

Insoo Hyun and Alejandro De Los Angeles (eds.), *Chimera Research: Methods and Protocols*, Methods in Molecular Biology, vol. 2005, https://doi.org/10.1007/978-1-4939-9524-0_7,

gene therapy application [4–7], though human hematopoiesis is limited to mostly B and some myeloid cells. Recently, a *c-kit* mutation was introduced into NSG and this new humanized mouse strain, NBSGW or NSGW41, exhibits improved human cell engraftment in the absence of conditioning during xenotransplantation. NBSGW mice also support multi-lineage reconstitutions (including human erythroid precursors) in the bone marrow and make it ideal for testing the efficacy of human HSC gene therapy vectors [2, 4, 5].

2 Materials

1. Cells: G-CSF mobilized peripheral blood leukopaks from healthy donors were purchased from Key Biologics, LLC (Memphis, TN), and CD34+ cells were purified using CliniMACS (Miltenyi Biotec, Auburn, CA) in the Human Application Lab at St. Jude Children's Research Hospital. The purified CD34+ cells were cryopreserved and stored in the vapor phase of liquid nitrogen for later use. Detailed procedures to thaw frozen human CD34+ cells are provided in Subheading 3.
2. Mouse: a. NBSGW strain (NOD.Cg-$Kit^{W\text{-}41J}$ Tyr^{+} $Prkdc^{scid}$ $Il2rg^{tm1Wjl}$/ThomJ) (JAX (stock 026622 https://www.jax.org/strain/026622)). No conditioning is required. The animals at 6–8 weeks of age are used as recipients for human CD34+ cells. b. Appropriate animal restrainers for tail vein injection and heat lamp to warm up mouse tails.
3. Lentiviral vectors: Lentiviral vectors were produced in 293 T cells using 4 plasmid transient transfection system and purified using ion exchange followed by diafiltration [8].
4. Recombinant human SCF, Recombinant human TPO, Recombinant human Flt3-Ligand (Peprotech, Rocky Hill, NJ).
5. X-VIVO 10 Chemically Defined, Serum-free Hematopoietic Cell Medium (Lonza, Walkersville, MD).
6. Corning® Penicillin-Streptomycin Solution, Corning® glutagro™, Supplement, 100×.
7. Protamine sulfate.
8. Iscove's Modified Dulbecco's Medium, MethoCult™ H4434 Classic (STEMCELL Technologies Inc., Cambridge, MA).
9. Falcon® 6, 12, 24-well Clear Flat Bottom Not Treated Cell Multiwell Culture Plate, 60 mm × 15 mm suspension culture dish, 100 mm × 20 mm non-treated culture dishes, Falcon™ Test Tube with Cell Strainer Snap Cap (Corning, NY).
10. 35 mm Culture dish with 2 mm grid.

11. Monoject 19 gauge blunt-end needles with 3 mL luer-lock syringe (Covidien (Medtronic), Minneapolis, MN).
12. Antibodies: Human CD3, CD19, CD33, CD45, CD235a, and mouse CD45, Ter119 (BD Biosciences). Fluorochrome conjugations are varied depending on the flow cytometry equipment and antibodies of choice. autoMACS Running Buffer—MACS Separation Buffer (Miltenyi Biotec, Auburn, CA) is used for staining and washing the samples.
13. Flow Cytometric Analysis

    In our facility, BD Biosciences LSR/Fortessa (San Jose, CA) and Attune NxT Flow Cytometer (Thermo Fisher Scientific, Grand Island, NY) are used for collecting flow cytometry data. FlowJo v.10 program (FlowJo, LLC, Ashland, OR) is used for the analysis.
14. Vector Copy Number Determination

    ddPCR Supermix for Probes (No dUTP), Automated Droplet Generator, C1000 Touch Thermal Cycler, QX200™ Droplet Digital™ PCR system with QuantaSoft Software (Bio-Rad), and QuantaSoft Analysis Pro (Bio-Rad, Hercules, CA).

3 Methods

3.1 The Overall Procedure Is Depicted in Fig. 1 (Cell Culture)

1. Prepare CD34+ culture media. Add the following cytokines and additives into X-VIVO 10 media: Human Flt3L (100 ng/mL), Human SCF (100 ng/mL), Human TPO (100 ng/mL), Pen/Strep at the final concentration of 50 U/mL and 50 μg/mL, 2 mM of L-ala L-glu (stock is 200 mM).
2. Thaw frozen CD34+ cells. Ensure that culture media is at room temperature.
3. Quickly thaw the frozen vial in a 37 °C water bath by gentle swirling until the last crystal piece is left (95% of ice has melted). This should take approximately 3 min (DO NOT allow the sample to warm to 37 °C). Cryovials should be cool to touch when removed from the water bath. Be careful not to submerge the entire vial in the water bath (submerge up to bottom of vial cap).
4. Wipe the outside of the vial with 70% ethanol and place it in a laminar flow hood.
5. Quickly transfer all of the cells from the cryovial to a 15 mL conical tube. If the cells are clumped, break them up (clumping can be eliminated by pipetting gently up and down as necessary).

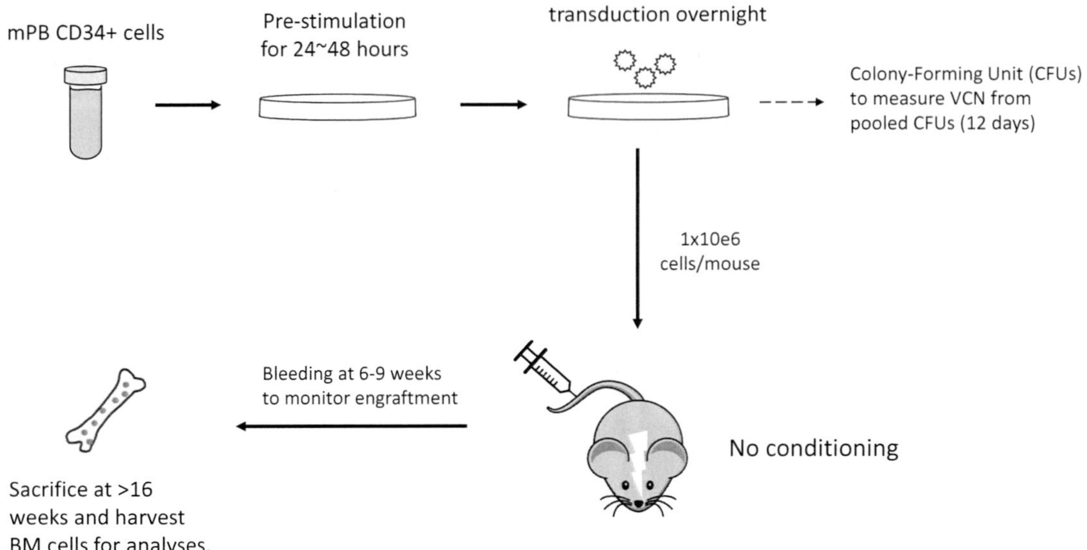

Fig. 1 Schematic diagram of human HSC transplantation into NBSGW mice to study lentiviral gene therapy

6. Then add 2 mL of cell culture media dropwise over a period of 2 min. Shake the tube using a vortex set to an appropriate level. Then add 8 mL of media in 1 mL aliquots slowly while vortexing. It is important to achieve a dilution ratio of 1:10 (sample to media).
7. Allow the cells to rest at room temperature for 35 min. Lay the tube horizontally so cells do not form a pellet.
8. The cells are centrifuged at 340 × *g* for 5 min in a large tabletop centrifuge. The supernatant is removed, and the cells are gently resuspended in fresh CD34+ culture media at 1 × 10E6 cells/mL for pre-stimulation.

3.2 Transduction

1. Pre-stimulation: At the 1 × 10E6 cells/mL with fresh media with cytokines as described above, cells are plated on 60 mm, 100 mm, or 150 mm dish (non-tissue culture treated) depending on the total volume. For example, for ten million CD34+ cells, use 100 mm dish with 10 mL volume that gives 1 × 10 E6 cells/mL. Incubate the cells at 37 °C in 5% CO_2 with ≥95% humidity for 24–48 h without disturbing cells.
2. After pre-stimulation, count the cells and transfer to a dish of appropriate volume at a cell density of two million cells per ml. For example, when using a 6-well plate, 2 mL media is used for a well and total cell number should be four million cells/well. Use non-tissue culture-treated plate without retronectin (*see* **Note 1**).
3. Add protamine sulfate at the final 8 μg/mL.

4. Add lentiviral vector at the multicity of infection (MOI) of 50. For example, in a well of 6-well with 2 mL culture volume, there are 4 × 10E6 total cells. At the MOI of 50, 2 × 10E8 transduction units (TU) would be required for the transduction (*see* **Note 2**).
5. Transduce overnight in 5% CO_2 incubator at 37 °C.

3.3 Transplantation into NBSGW Recipient Mice

1. 16–24 h later, cells are collected into an appropriate tube and wash the plate with the media. Centrifuge the tube at 490 × *g* for 5 min, and resuspend the pellet with the CD34+ culture media with cytokines.
2. Count cells and use 1800 cells (see the CFU assay section below) for colony-forming unit (CFU) assays and centrifuge again at 490 × *g* for 5 min.
3. Resuspend cells in an appropriate volume of PBS with 2% FBS and Pen/Strep or Iscove's Modified Dulbecco's Medium (IMDM) for transplantation (*see* **Note 3**).
4. NBSGW mice do not require any conditioning prior to transplantation. If using NSG mice, *see* **Note 4**.
5. Tail vein injection: Put the mice under a heat lamp in a cage to warm up the tail and wait. See if the vein is dilated. Mouse lateral vein on either side is used for the intravenous injection.
6. Once the vein is dilated, put the mouse onto an appropriate mouse restrainer (*see* **Note 5**).
7. Wipe the injection site on the tail with 70% alcohol (with a cotton swab) before the injection.
8. The solution containing the transduced human cells is infused using a 1 mL syringe with 27 G needle. Avoid any bubble as much as possible. Hold the mouse tail tight.
9. With the bevel of the needle facing upward and the needle almost parallel to the vein, slide the needle into the tail vein. Confirm the location by seeing a flash of blood on the needle. There should be no resistance and the tail vein becomes pale and turns back to red immediately after the injection when the injection is done properly.
10. Remove the needle from the vein and apply slight pressure to the puncture site with a dry piece of gauze.
11. Place the animal back to a clean cage and monitor the animal for a minute.
12. Antibiotics after transplantation: Add Baytril to water for 3 weeks (the stock concentration is 2.27% or 22.7 mg/mL). Add 90.8 mg (4 mL) into a 350 mL bottle of water. Then, alternate Baytril with Sulfatrim every other week (Sulfameth oxazole and Trimethoprim oral suspension, 200 mg/40 mg per 5 mL). Add 7 mL into a 350 mL bottle of water.

3.4 Colony-Forming Unit (CFU) Assays to Determine VCN

1. Aliquot 3 mL of methylcellulose medium (MethoCult™ H4434 Classic, Catalog # 04434) or use pre-aliquoted media following the manufacturer's instructions with some modification as follows.
2. Add 1800 cells (which were transduced and saved from the culture as previously described) to 3 mL methylcellulose medium and vortex the tube vigorously for 5 s to mix well. Let stand for 5 min.
3. Attach a Monoject 19 G blunt-end needle to a 3 mL luer-lock syringe. Aspirate the methylcellulose and split into two of 35 mm plates with grid. Try to avoid making bubbles as much as possible. Distribute the medium evenly across the surface by gently tilting and rotating the plate (*see* **Note 6**).
4. Place the culture plates into the outer dish (for example 150 mm petri dish). Add approximately 3 mL of sterile water to an uncovered 35 mm plate and place it together with those MethoCult plates. Place the lid onto the outer dish.
5. Incubate plates in 37 °C CO_2 incubator for 12–14 days.
6. Colonies are collected at 12 days by scraping with 1 mL pipette tips or serological pipettes once 2–3 mL of PBS is added onto the plates. Use more PBS to collect as many cells as possible by rinsing the dishes 2–3 times and pool those cells into 15 mL conical tubes. Then, centrifuge at 490 × *g* for 5 min.
7. Cells are washed with 10 mL PBS and collected by centrifugation again.
8. Genomic DNA (gDNA) from the pool colonies is extracted using Quick-DNA Microprep kit (Zymo Research, Irvine, CA) following the manufacturer's instructions but increasing the centrifugation time to 3–5 min and eluting in 30 μL of elution buffer.
9. Then 16 μL of gDNA is digested at 37 °C for ≥45 min using 2 μL of MspI enzyme and 2 μL of CutSmart Buffer (New England Biolabs).
10. At the end of the incubation time, digested gDNA is diluted 1:5 adding 80 μL of nuclease-free water and an additional dilution 1:3 is performed (20 μL of diluted gDNA and 40 μL nuclease-free water). Both dilutions are tested in ddPCR.
11. To evaluate VCN, ddPCR is performed using a primer-probe set recognizing the lentiviral vector packaging signal (ψ); RPP30 is selected as a reference gene. Primers and probes used were synthesized by Integrated DNA Technologies.
12. PCR reactions are set up to detect concurrently psi and RPP30 as follows: 12.5 μL of ddPCR™ Supermix for Probes No dUTP (Bio-Rad Laboratories, Hercules, CA), 0.225 μL of each primer (final concentration 900 nM), 0.0625 μL of each

probe (final concentration 250 nM), 5.225 μL of nuclease-free water, and 6.25 μL of diluted gDNA.

13. Automated Droplet Generator (Bio-Rad Laboratories) and appropriate consumables are used to generate the droplets starting from the reactions. The ddPCR is run on a C1000 Touch Thermal Cycler (Bio-Rad Laboratories) using the following protocol: enzyme activation 95 °C for 10 min, 40 cycles of denaturation at 94 °C for 30 s (ramp rate 2 °C/s) and annealing/extension at 60 °C (ramp rate 2 °C/s) for 1 min, enzyme deactivation 10 min at 98 °C, and hold at 4 °C.
14. Droplets are evaluated using a QX200 Droplet Digital PCR System (Bio-Rad Laboratories) with QuantaSoft Software (Bio-Rad Laboratories).
15. Droplet analysis and copy number calculation are performed using QuantaSoft Analysis Pro (Bio-Rad Laboratories) considering the reference gene as two copies (since diploid cells are analyzed). The average of the VCN of the two dilutions tested is then reported.

3.5 Flow Cytometric Analyses for the Engraftment and the Expression of Lineage Markers

1. To monitor the human cell engraftment, collect mouse peripheral blood samples once at 6–10 weeks by retro-orbital bleeding with capillary tubes under anesthesia with isoflurane and transfer to Eppendorf tubes (these capillary tubes are coated with either heparin or EDTA). The engraftment level in the bone marrow is correlated with the human CD45+ level in the peripheral blood. Typically, 3–25% of leukocytes are human at 8–12 weeks posttransplantation.
2. For the final analyses, sacrifice the mice at >16 weeks posttransplantation and collect two humeri, two femurs, two tibias, and two iliac crests from each mouse (*see* **Note 7**).
3. Position mouse on its back and wet fur thoroughly with 70% isopropyl alcohol. This step decreases the possibility of contaminating cell preparations with fur. Pin upper limbs.
4. Cut a slit in the fur just below the rib cage without cutting the peritoneal membrane. Non-sterile scissors can be used for this step.
5. Firmly grasp skin and peel back to expose hind limbs. Using sterile sharp dissecting scissors, cut the knee joint in the center. Cut through ligaments and excess tissue. Use of sharp scissors will prevent splitting of the bone.
6. Grasp femur with forceps and cut femur near hip joint.
7. Free tibia by cutting near the ankle joint. Trim the ends of the long bones to expose the interior marrow shaft.
8. Prepare iliac crests (hip bones) and humerus in a similar way.
9. Put bones in a sterile petri dish or in sterile culture medium and place on ice.

10. Use a 5 mL syringe with a 25 G needle, and draw up 5 mL of cold PBS containing 2% heat-inactivated (HI) FBS and Pen/-Strep. A smaller needle (27 G) may be required to remove cells from the tibia, humerus, or iliac crests.
11. Cells should be isolated as soon as possible after the animal is sacrificed. Insert bevel of the needle into marrow shaft and flush marrow into sterile 50 mL conical tube. Insert cell strainer (100 μm) on the tube to avoid debris and clumping and let the cells pass through the strainer. The bone should appear white once all the marrow has been expelled.
12. Keep the cells on ice until use.
13. Centrifuge at 490 × *g* for 5 min.
14. Aspirate the solution without touching red smear/pellet and gently resuspend in PBS containing 2% HI FBS and Pen/Strep or autoMACS buffer for further staining procedures.
15. Perform cell counting and staining with fluorescent antibodies as follows: human CD3 for T cells, CD19 for B cells, CD33 for myeloid cells, human CD45 for human hematopoietic lineage cells, hCD235a for erythroid cells, mouse CD45 for mouse hematopoietic lineage cells, and mouse Ter119 for mouse erythroid cells. DAPI is used as a dead cell marker (*see* **Note 8**). Examples of flow analyses are shown in Fig. 2. Engraftment of human cells is more than 90% at 16 weeks and abundant human erythroid hCD235a^{+} cells are also present.

4 Notes

1. For lentiviral transduction, retronectin-coated plates are not necessary as transduction efficiency is not much improved with retronectin treatment, unlike retroviral vectors. Therefore, we use non-tissue culture-treated plates.
2. Lentiviral vectors are purified and formulated in the X-VIVO 10 before freezing. A small aliquot of the frozen vector should be thawed and titer is determined before transduction.
3. IMDM gives more visibility of injecting the cells. We usually use 200 μL per injection (i.e., per mouse) which contains around a million cells. Always have some extra volume for resuspension, e.g., when preparing for five mice, use 1.1 mL instead of 1.0 mL considering dead spaces of needle/syringe, and bubbles.
4. NSG strain (NOD.Cg-*Prkdc*^{scid} *Il2rg*^{tm1Wjl}/SzJ, also known as NOD scid gamma).

   These humanized mice are available at JAX (stock 005557 https://www.jax.org/strain/005557). Typically, 6- to

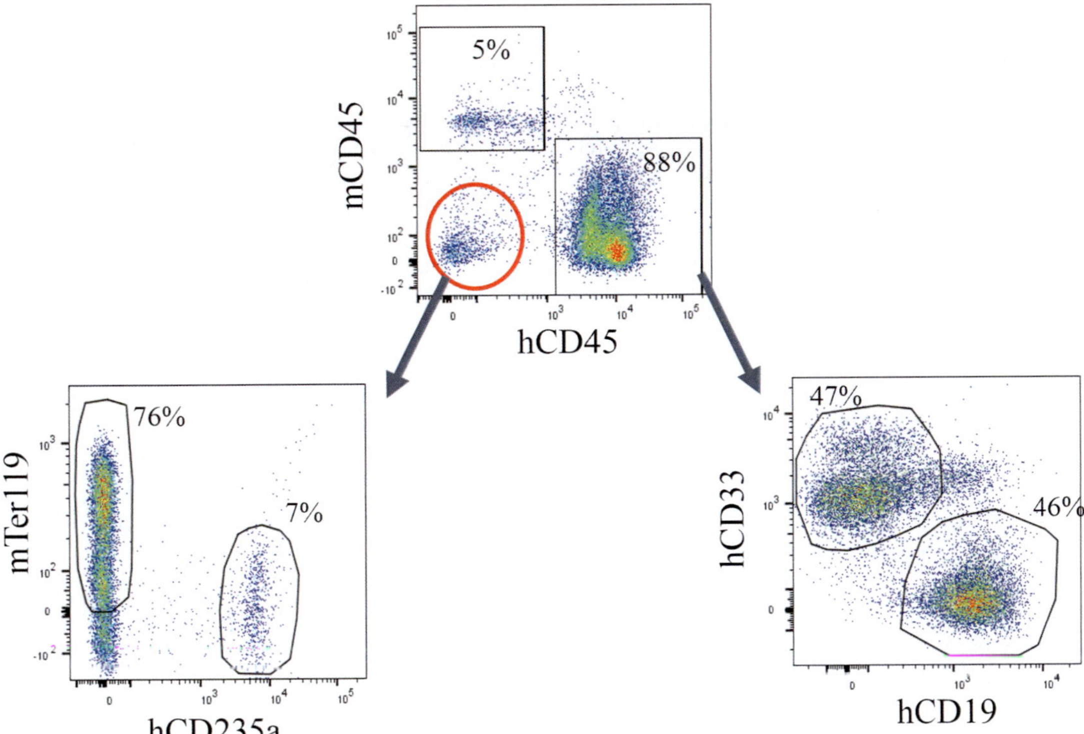

Fig. 2 A representative flow cytometry gating strategy for the analysis of expression of lineage markers: hCD19 for B lymphoid cells, hCD33 for myeloid cells. These data were obtained from NBSGW mouse BM transplanted with mobilized human peripheral blood CD34+ cells at 16 weeks posttransplantation

12-week-old mice are used for transplantation. Conditioning is required for xenotransplantation: 24 h prior to transplantation using either Busulfan at 35 mg/kg or irradiation at 150–200 rads. All the conditioning must be done 1 day prior to transplantation. Baytril-treated water should be put on the mouse cages as well at the conditioning as a preventive measure.

5. Try not to overheat the mouse tail as it could cause vasoconstriction eventually and hyperthermia/burn-related injuries to the animals.
6. CFU assays should be done in a sterile way. Use a separate needle and a syringe for each sample.
7. In our case, we need to obtain as many cells as possible to sort the cells for our purpose. Therefore, we collect cells from those bones as listed. However, if you do not need to sort the cells, femurs and tibias are enough for the collection of bone marrow cells.
8. Different combinations of the fluorochrome-conjugated antibodies can be selected depending on flow cytometer laser layout.

References

1. Shultz LD, Lyons BL, Burzenski LM, Gott B, Chen X, Chaleff S, Kotb M, Gillies SD, King M, Mangada J, Greiner DL, Handgretinger R (2005) Human lymphoid and myeloid cell development in NOD/LtSz-scid IL2R gamma null mice engrafted with mobilized human hemopoietic stem cells. J Immunol 174 (10):6477–6489
2. Shultz LD, Brehm MA, Garcia-Martinez JV, Greiner DL (2012) Humanized mice for immune system investigation: progress, promise and challenges. Nat Rev Immunol 12(11):786–798. https://doi.org/10.1038/nri3311
3. Greiner DL, Hesselton RA, Shultz LD (1998) SCID mouse models of human stem cell engraftment. Stem Cells 16(3):166–177. https://doi.org/10.1002/stem.160166
4. McIntosh BE, Brown ME, Duffin BM, Maufort JP, Vereide DT, Slukvin II, Thomson JA (2015) Nonirradiated NOD,B6.SCID Il2rgamma−/− Kit(W41/W41) (NBSGW) mice support multilineage engraftment of human hematopoietic cells. Stem Cell Reports 4(2):171–180. https://doi.org/10.1016/j.stemcr.2014.12.005
5. Rahmig S, Kronstein-Wiedemann R, Fohgrub J, Kronstein N, Nevmerzhitskaya A, Bornhauser M, Gassmann M, Platz A, Ordemann R, Tonn T, Waskow C (2016) Improved human erythropoiesis and platelet formation in humanized NSGW41 mice. Stem Cell Reports 7(4):591–601. https://doi.org/10.1016/j.stemcr.2016.08.005
6. Saito Y, Ellegast JM, Rafiei A, Song Y, Kull D, Heikenwalder M, Rongvaux A, Halene S, Flavell RA, Manz MG (2016) Peripheral blood CD34(+) cells efficiently engraft human cytokine knock-in mice. Blood 128(14):1829–1833. https://doi.org/10.1182/blood-2015-10-676452
7. Walsh NC, Kenney LL, Jangalwe S, Aryee KE, Greiner DL, Brehm MA, Shultz LD (2017) Humanized mouse models of clinical disease. Annu Rev Pathol 12:187–215. https://doi.org/10.1146/annurev-pathol-052016-100332
8. Greene MR, Lockey T, Mehta PK, Kim YS, Eldridge PW, Gray JT, Sorrentino BP (2012) Transduction of human CD34+ repopulating cells with a self-inactivating lentiviral vector for SCID-X1 produced at clinical scale by a stable cell line. Hum Gene Ther Methods 23 (5):297–308. https://doi.org/10.1089/hgtb.2012.150

Chapter 8

Pig Chimeric Model with Human Pluripotent Stem Cells

Cuiqing Zhong, Jun Wu, and Juan Carlos Izpisua Belmonte

Abstract

Interspecies chimera formation provides a unique platform for studying donor cell developmental potential, modeling disease in vivo, as well as in vivo production of tissues and organs. The derivation of human pluripotent stem cells (hPSC) from either human embryos or somatic cell reprogramming facilitates our understanding of human development, as well as accelerates our exploration of regenerative medicine for human health. Due to similar organ size, close anatomy, and physiology between pig and human, human-Pig interspecies chimeric model in which pig serves as the host species may open new avenues for studying human embryogenesis, disease pathogenesis, and generation of human organ for transplantation to solve the worldwide donor organ shortage. Our previous study demonstrated chimeric competency of different types of human PSCs in pig host. In this chapter, we introduce our workflow for the generation of human PSCs and analysis of its chimeric contribution to pre- and postimplantation pig embryos.

Key words Human pluripotent stem cells (hPSCs), Human embryonic stem cells (hESCs), Human induced pluripotent stem cells (hiPSCs), Blastocyst, Epiblast stem cells (EpiSCs), Pig, Chimeric contribution

1 Introduction

Pluripotent stem cells (PSCs) are characterized by their ability to generate all cell types of an adult organism, which can be obtained from preimplantation embryos or somatic nuclear reprogramming [1–4]. The development of different culture recipes has led to the stabilization of discrete pluripotent states in vitro reflective of different phases of pluripotency in epiblast cells in vivo. Most recognized pluripotent states include naïve and primed states. Cells in the naïve state possess unbiased developmental potential both in vivo and in vitro while primed cells show more restricted developmental potential and are primed for lineage differentiation [5–10]. In rodents, when naïve embryonic stem cells (ESCs) are introduced into the host blastocyst, they could extensively contribute to a wide range of tissues and organs. In contrast, primed epiblast stem cells (EpiSCs) are incapable for chimeric formation in blastocyst due to unmatched developmental timing [8, 11].

Insoo Hyun and Alejandro De Los Angeles (eds.), *Chimera Research: Methods and Protocols*, Methods in Molecular Biology, vol. 2005, https://doi.org/10.1007/978-1-4939-9524-0_8,

The successful derivation of human pluripotent stem cells (hPSCs), including human embryonic stem cells (hESCs) and human induced pluripotent stem cells (hiPSCs), revolutionizes our studies of human development and regenerative medicine. Despite their blastocyst origin, hESCs maintained in conventional culture medium reside in primed state and share many defining features of rodent EpiSCs [6, 8]. In recent years, a number of studies reported new culture conditions for the generation and long-term maintenance of naïve-like hPSCs [12–14]. These naïve-like hPSCs were shown to share molecular similarities to mESCs; however, their in vivo developmental potential remains to be determined. Another putative advantage of obtaining true naïve hPSCs is the prospect of generating functional human tissue and organ in vivo via interspecies blastocyst complementation [15].

The pig has emerged as one of the most popular large animal models in biomedical research, largely due to the similarity in organ size, physiology, development, and disease progression to humans. Our previous study has demonstrated the possibility of using pig as an in vivo model for studying hPSC developmental potential via the formation of human-pig chimeric embryos [16]. The study of pig chimeric models with hPSCs may pave the way for human organ generation, in vivo new drug testing, and human disease modeling [15].

In this chapter, we will provide detailed protocols for the generation of hiPSCs, chimera formation with hiPSCs, and characterization of human-pig interspecies chimeric embryos at pre- and postimplantation stages (Fig. 1). There are many reported culture conditions for hPSCs, which presumably stabilize hPSCs at distinct pluripotent states. Here, we focus on two culture conditions

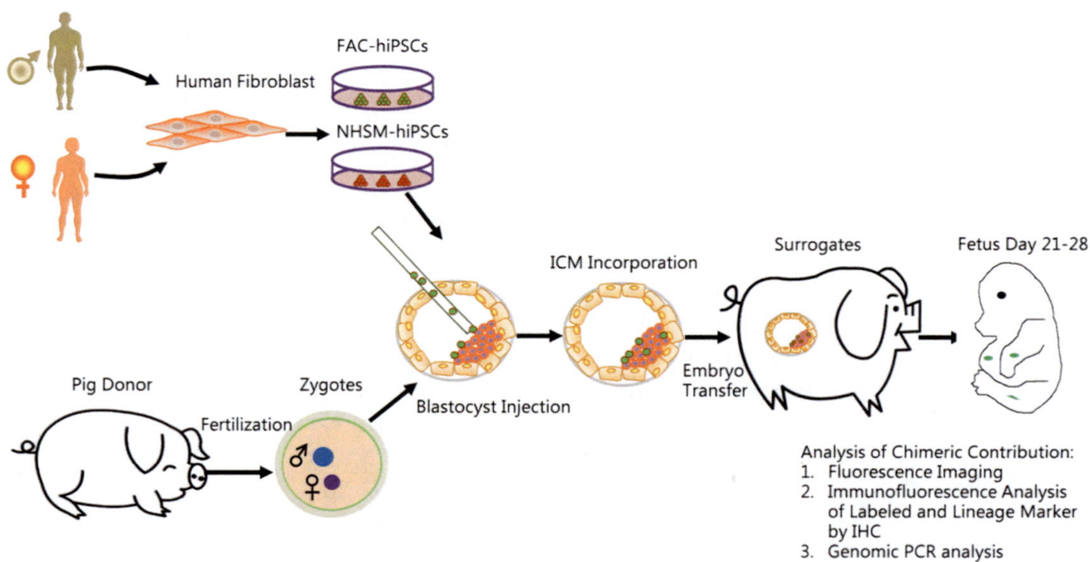

Fig. 1 Schematic representation of strategy for human-pig interspecies chimeras

including NHSM condition that can arrest human cells at a naïve-like pluripotent state [12] and also the FAC condition that enables the generation of intermediate hiPSCs presumably between naïve and primed pluripotent states [16, 17].

2 Materials

2.1 Generation and Characterization of Human Induced Pluripotent Stem Cell

2.1.1 Generation of Naïve-Like and Intermediate hiPSCs

1. Human foreskin fibroblast (HFF).
2. 0.1% (w/v) Gelatin (GIBCO).
3. Centrifuge (Thermo Fisher).
4. Falcon tube (15 and 50 mL).
5. Mitotically inactivated mouse embryonic fibroblast (MEF) and puromycin-resistant DR4 MEFs.
6. MEF medium: DMEM (GIBCO), 10% fetal bovine serum (FBS) (GIBCO), 0.1 mM nonessential amino acid (NEAA) (GIBCO), 1× Pen-Strep (GIBCO).
7. Phosphate-buffered saline (PBS) without Ca^{2+} and Mg^{2+} (Corning).
8. TrypLE (GIBCO).
9. Accumax (Innovative Cell Technologies).
10. P2 Primary Cell 4D-Nucleofector Kit (Lonza) for HFF.
11. 4D-Nucleofector with X-Unit (Lonza)
12. NHSM medium: To prepare 500 mL of NHSM medium we added 500 mL KnockOut DMEM (GIBCO), 5 mL Pen-Strep (GIBCO), 5 mL GlutaMax (GIBCO), 5 mL NEAA (GIBCO), 2.5 g AlbumaxI (GIBCO), 5 mL N2 supplement (GIBCO), 50 μg/mL ascorbic acid 2-phosphate (Sigma), 20 ng/mL human LIF (Peprotech), 20 ng/mL human LR3-IGF1 (Peprotech), 8 ng/mL FGF2 (Peprotech), 2 ng/mL TGFβ1 (Peprotech), 3 μM CHIR99021 (Selleckchem), 1 μM PD0325901 (Selleckchem), 5 μM SB203580 (Selleckchem), 5 μM SP600125 (Selleckchem), 5 μM Y27632 (Tocris), and 0.4 μM LDN193189 (Selleckchem). Filter with 0.22 μm filter.
13. FAC medium: To prepare 500 mL of FAC medium, 250 mL DMEM/F12 (GIBCO) and 250 mL neurobasal medium (GIBCO) were mixed at 1:1 ratio, 2.5 mL N2 supplement (GIBCO), 5 mL B27 supplement (GIBCO), 5 mL GlutaMax (GIBCO), 5 mL NEAA (GIBCO), 0.1 mM β-mercaptoethanol (GIBCO), 5 mL Pen-Strep (GIBCO), 50 μg/mL BSA (optional) (Sigma); 12 ng/mL FGF2 (Peprotech), 10 ng/mL activin-A (Peprotech), and 3 μM CHIR99021. Filter with 0.22 μm filter.
14. Cryogenic vials (NALGENE).

2.1.2 Immunostaining

1. PBST: PBS containing 0.1% Tween-20 (Fisher Scientific).
2. Permeabilization solution: PBS containing 0.3% Triton X-100 (Fisher Scientific).
3. Blocking solution: PBST containing 3% (w/v) BSA (Sigma).
4. Primary antibodies: OCT4 (Santa Cruz), NANOG (R&D system), TRA-1-60 (Santa Cruz), and TRA-1-80 (Santa Cruz).
5. Fluorescence-conjugated secondary antibodies (Alexa Fluor antibodies, Thermo Fisher Scientific).
6. ProLong Gold Antifade Mountant (Thermo Fisher Scientific).
7. Inverted fluorescence microscope.
8. Coverslip.

2.1.3 Karyotyping

1. Demecolcine (Sigma).
2. Hypotonic solution: Dissolve 0.25 g sodium citrate and 0.28 g KCl in 100 mL dH_2O. Store at RT.
3. Fixative solution: Mix methanol and acetic acid at a ratio of 3:1 (in volume). Freshly prepare before use and keep in −20 °C. Prepare the solution in a chemical fume hood.
4. 100% Ethanol.
5. Staining jar.
6. Trypsin without phenol red (GIBCO, powder).
7. Earle's Balanced Salt Solution (Thermo Fisher Scientific).
8. Hot metal plate or alcohol burner.
9. 10× Giemsa (GIBCO): To make 1× Giemsa staining solution, dilute 1 mL10× Giemsa with 9 mL PBS, and then filter with 0.22 μm filter.
10. Microscope with 100× oil objective.

2.1.4 Teratoma Assay

1. 29G insulin syringe (BD Bioscience).
2. NHSM or FAC culture medium.
3. Matrigel (BD Bioscience).
4. NOD-SCID mice.
5. Paraffin.
6. Hematoxylin and eosin.

2.1.5 Generation of Fluorescently Labeled hiPSCs

1. pCAG-IP-humanized Kusabira Orange (hKO).
2. pEGIP (addgene#27078).
3. Cell counter.
4. Puromycin (Invivogen).
5. P3 Primary Cell 4D-Nucleofector Kit (Lonza) for hiPSCs.

2.2 Interspecies Human-Pig Chimera Formation

2.2.1 Preparation of Pig Blastocysts

Generation of Pig Blastocysts Through Parthenogenesis

1. TCM-199 (GIBCO).
2. Porcine follicular fluid stock (10×): Aspirate and collect follicular fluid from ovaries at 3–6 mm of follicle. Centrifuge at 1200 × g for 60 min at 4 °C. Filter, aliquot, and store at −20 °C.
3. PBS-PVA: PBS containing 0.1% (w/v) polyvinyl-alcohol (PVA). To prevent contamination, add penicillin and streptomycin.
4. TL-HEPES-PVA: HEPES-buffered Tyrode's lactose medium with 0.1% (w/v) PVA. 113.8 mM NaCl, 3.2 mM KCl, 2 mM $NaHCO_3$, 0.35 mM NaH_2PO_4, 10 mM sodium lactate, 2 mM CaCl2, 0.5 mM $MgCl_2$, 10 mM HEPES, 0.027 mM phenol red, 5 mM glucose, 43.9 mM sorbitol, 50 mg/L streptomycin, 65 mg/L penicillin in water. Adjust pH to 7.2–7.4 with NaOH. Use 0.22 μm filter to sterilize. Store at 4 °C. Warm the medium before use. TL-HEPES-PVA can be replaced by HEPES-buffered TCM-199 plus with 10% FCS, and here we also use TCM-199 to culture the embryos.
5. In vitro maturation medium (IVM): TCM-199 containing 0.1% (w/v) VPA, 3.05 mM D-glucose, 0.91 mM sodium pyruvate, 0.5 μg/mL oFSH, 0.5 μg/mL bLH, 10 ng/mL EGF, 10 μg/mL gentamicin (GIBCO), and 10% (v/v) porcine follicle fluid.
6. MEM: Minimum essential medium (GIBCO).
7. HEPES (GIBCO).
8. BTX Electro Cell Manipulator 2001 (BTX).
9. PZM-5: Porcine Zygote Medium-5 (Research Institute for the Functional Peptides), medium for in vitro embryo culture.
10. Hyaluronidase: 1 mg/mL Solution in HEPES-TL-PVA.
11. Cytochalasin B: 5 mg/mL Stock solution in DMSO. Aliquots and freeze at −20 °C.
12. Thermo-container: To keep temperature during transportation of ovaries and embryos.
13. Incubator: CO_2 controlled (5% CO_2 in air) for cell culture and both CO_2 and O_2 controlled (5% CO_2, 5% O_2, and 90% N_2) at 38.5 °C for embryo culture.
14. Water bath.
15. Mineral oil.
16. Stereomicroscope with heating stage.
17. Mouth pipette.

Generation of Pig Blastocysts Through In Vivo Fertilization

1. Sows.
2. eCG: Equine chorionic gonadotropin (Merck Animal Health),
3. hCG: Human chorionic gonadotropin (Merck Animal Health),

4. Midventral laparotomy.
5. Azaperone (Stresnil).
6. Sodium thiopental.
7. Isoflurane.
8. TL-HEPES-PVA medium.
9. Thermostatically controlled incubator.
10. Glucose-free NCSU (North Carolina State University)-23 medium: 108.73 mM NaCl, 4.78 mM KCl, 1.70 mM $CaCl_2$, 1.19 mM KH_2PO_4, 1.19 mM $MgSO_4$, 25.07 mM $NaHCO_3$, 5.55 mM glucose, 7 mM taurine, 50 mg/L streptomycin, and 65 mg/L penicillin in water. Add L-glutamine (final concentration: 1mM) and hypotaurine (final concentration: 5 mM) before use. Use 0.22 μm filter to sterilize and store at 4 °C for up to 2 weeks.
11. Pyruvate (GIBCO).
12. Lactate (Sigma).
13. Nunc 4-well cell culture-treated multidish (Thermo Fisher Scientific).
14. Fetal calf serum.
15. Stereomicroscope with heating stage.

2.2.2 Microinjection of hiPSCs to Pig Blastocysts and Embryo Culture

1. Micromanipulator with a laser system (Saturn 5 Active, Research Instruments).
2. Micropipettes: 20–30 μm Internal diameter blunt-end micropipettes for cell injection. And 150 μm internal diameter blunt-end for holding pipette.
3. Inverted fluorescence microscope (Nikon, Tokyo).

2.2.3 Pig Embryo Transfer

1. Gynetics embryo transfer catheter (Gynetics Medical Products N.V).
2. Surgical suture or clamps for the wound.
3. NCSU-23 supplemented with 10 mM HEPES, 0.4% (v/v) BSA, and 10% (v/v) FCS.
4. Altrenogest.
5. Telazol (Zoetis, Kalamazoo).

2.2.4 Collection of E21–28 Embryos by Pig Surrogate Cesarean Section

1. Transabdominal ultrasonography (WED-2000AV, Welld).
2. Euthanasia solution (Fatal Plus Solution, Vortex Pharmaceutical Ltd).
3. Isoflurane.
4. Thermo container: For fetus transportation.

2.3 Analysis of Chimeric Contribution of hiPSC to Pig Blastocysts

2.3.1 Immunostaining of HuNu and SOX2 in Pig Blastocysts

1. Primary antibodies: HuNu (Millipore Cat#MAB1281) and SOX2 (BioGenex Cat# NU579-UC).
2. Paraformaldehyde: 4% (w/v) dissolved in PBS, and store at 4 °C.
3. PBS-PVA: PBS containing 1 mg/mL PVA.
4. Washing buffer: PBS-PVA containing 0.1% Triton X-100.
5. Fluorescent-conjugated second antibodies (Alexa Fluor antibodies, Thermo Fisher Scientific).
6. ProLong Gold Antifade Mountant (Thermo Fisher Scientific).

2.3.2 Evaluation of Chimeric Efficiency in Blastocyst

1. Inverted fluorescence microscope.

2.4 Analysis of Chimeric Contribution of hiPSC to Postimplantation Pig Embryos

2.4.1 Evaluation of Pig Embryo Developmental Status

1. Balance.
2. Stereoscope.

2.4.2 Fluorescence Imaging

1. Epifluorescence stereomicroscope.

2.4.3 IHC Analysis of Fluorescence Signal and Lineage Markers

1. 30% (w/v) Sucrose solution (Sigma): Dissolve in water.
2. OCT (Sakura Finetek).
3. Dry ice.
4. Antigen retrieval solution (HistoVT one, Nacalai Tesque).
5. Primary antibody: Rabbit anti-monomeric Kusabira-Orange 2 (MBL Code #PM05 1 M 1:500), Rabbit anti-GFP (MBL Code # 598, 1:500), Rat anti-CK8 (TROMA-1, DSHB Antibody Registry ID: AB_531826, 1:20), Mouse monoclonal anti-Ep-CAM (Santa Cruz Cat# sc-25308, 1:50), Mouse anti-actin α-smooth muscle (Sigma Cat# A5228, 1:200), Mouse anti-tubulin β3 (Clone TUJ1) (Biolegend Cat#801202, 1:500).

2.4.4 Genomic PCR

1. DNeasy Blood and Tissue Kit (QIAGEN).
2. PrimeSTAR GXL DNA Polymerase (Takara).
3. 0.2 mL PCR tube.
4. Thermal cycler.

5. Agarose.
6. Electrophoresis system.

3 Methods

3.1 Generation and Characterization of Human Induced Pluripotent Stem Cell

3.1.1 Generation of Naïve-Like and Intermediate hiPSCs

1. Prepare 6-well plate with 2 mL 0.1% gelatin in each well, and incubate at 37 °C for more than 30 min. Thaw one vial of low-passaged human foreskin fibroblasts (HFF) from liquid nitrogen, and place in 37 °C water bath immediately. Swirl gently for 1–2 min and then transfer the cells into 15 mL Falcon tube containing 10 mL pre-warmed MEF medium. Centrifuge at 200 × *g* for 5 min and discard the medium. Resuspend the pellet into 2 wells of 6-well plate culturing with MEF medium. Rock the dishes back and forth for a few times to make the cells disperse evenly. Place the plates into a humidified cell culture incubator. Change medium every two days. After cells reaching ~80% confluence, passage cells into one 100 mm dish.
2. Once the HFF reaches 90% confluence, aspirate the medium, and wash the cell with 5 mL PBS. Dissociate the cells with 1 mL TrypLE. Incubate the cell at 37 °C until individual cells start to round up. Add 5 mL MEF medium and dissociate cells with repeated pipetting. Collect 2×10^6 cells in one 15 mL Falcon tube (*see* **Note 1**). Centrifuge at 200 × *g* for 5 min and discard the medium. Rinse once with PBS.
3. Resuspend the cell pellet with the Nucleofection/Plasmid mixture (P2 Primary Cell 4D-Nucleofector Kit, Lonza): 82 μL Solution +18 μL supplement1 + 2 μg of each plasmid pCXLE-EGFP, pCXLE-hOCT3/4-shp53-F, pCXLE-hSK, pCXLE-hUL (*see* **Note 2**).
4. Put the cells into cuvette and perform nucleofection using 4D-Nucleofector following the protocol recommended by the manufacturer (Program EN150, Lonza). Add 500 μL MEF medium to cuvette after nucleofection, resuspend, and seed the cells into one 100 mm dish. Change the medium every 2 days.
5. Five days post-nucleofection, dissociate the HFF with TrypLE and replate the cells onto fresh prepared mitotically inactivated MEFs (*see* **Note 3**). Rock the dishes back and forth for a few times to make the cells disperse evenly. Change the medium to NHSM medium (for naïve-like hiPSCs) or FAC medium (for intermediate hiPSCs) the next day and continue culturing the cells. Change medium every 2 days initially and then every day when the cells become ~70% confluent. Check the colony formation during cell reprogramming under microscope.

6. After 20–30 days, pick individual hiPSC colonies and transfer to newly prepared MEFs in 24-well plate for further cultivation and expansion (*see* **Note 4**). After 4–8 days' culturing, passage the hiPSC colonies when they become big enough (Fig. 2a). Rinse the cell with PBS once, and add 150 μL Accumax into each well. Incubate at 37 °C for 1–2 min. Detach the cells from plate by repeated pipetting with pre-warmed medium. Collect the cell in Falcon tube and centrifuge. After aspirating the supernatant, distribute the cell pellet into 6-well plate in 2–3 mL NHSM or FAC medium. hiPSCs were passaged every 4–6 days in single cell using Accumax in the presence of 10 μM Y-27632 (*see* **Note 5**). Change medium every day.
7. To freeze the hiPSC, dissociate with Accumax and collect the cells into 15 mL Falcon tube. Centrifuge at 200 × *g* for 5 min and discard the medium. Resuspend the cell pellet with cold (4 °C) freezing medium, and then transfer the cell into cryogenic vials (1 mL per vial). Always freeze 1–2 × 10^6 cells in each vial or cell with 80% confluence from 1 well of 6-well plate. Place the cells on ice for 10–15 min, transfer cell vials into an isopropanol chamber, and place the chamber into −80 °C overnight. The next day, transfer the cell vials into liquid nitrogen.
8. To thaw the hiPSCs, follow **step 1**, and plate the cells into prepared fresh MEF plate with 2–3 mL NHSM or FAC medium containing 10 μM Y-27632 in each well of 6-well plate. Change the medium every day.

3.1.2 Immunostaining

1. Prior to the experiment, seed hiPSCs into 24-well plate with coverslip.
2. 2–4 Days later, aspirate the medium and wash the cells once with PBS. Take out the coverslips using a pair of forceps with care, and then fix the cells by incubating in 4% PFA solution at RT for 20 min. Wash the coverslips three times in washing solution PBST at RT for 10 min each.
3. Permeabilize the cells in permeabilization solution at RT for 20 min. Wash the cells three times in PBST at RT for 10 min each.
4. Incubate the cells in blocking solution at RT for 1 h.
5. Incubate the cells with primary antibodies OCT4, NANOG, TRA-1-60, and TRA-1-80 diluted at 1:200, 1:200, 1:200, and 1:200 in blocking solution overnight at 4 °C. After that, wash the cells three times in PBST at RT for 10 min each.
6. Incubate the cells with fluorescence-conjugated secondary antibodies diluted 1:500–1:1000 in blocking solution at RT for 1 h. Wash the cells three times in PBST at RT for 10 min

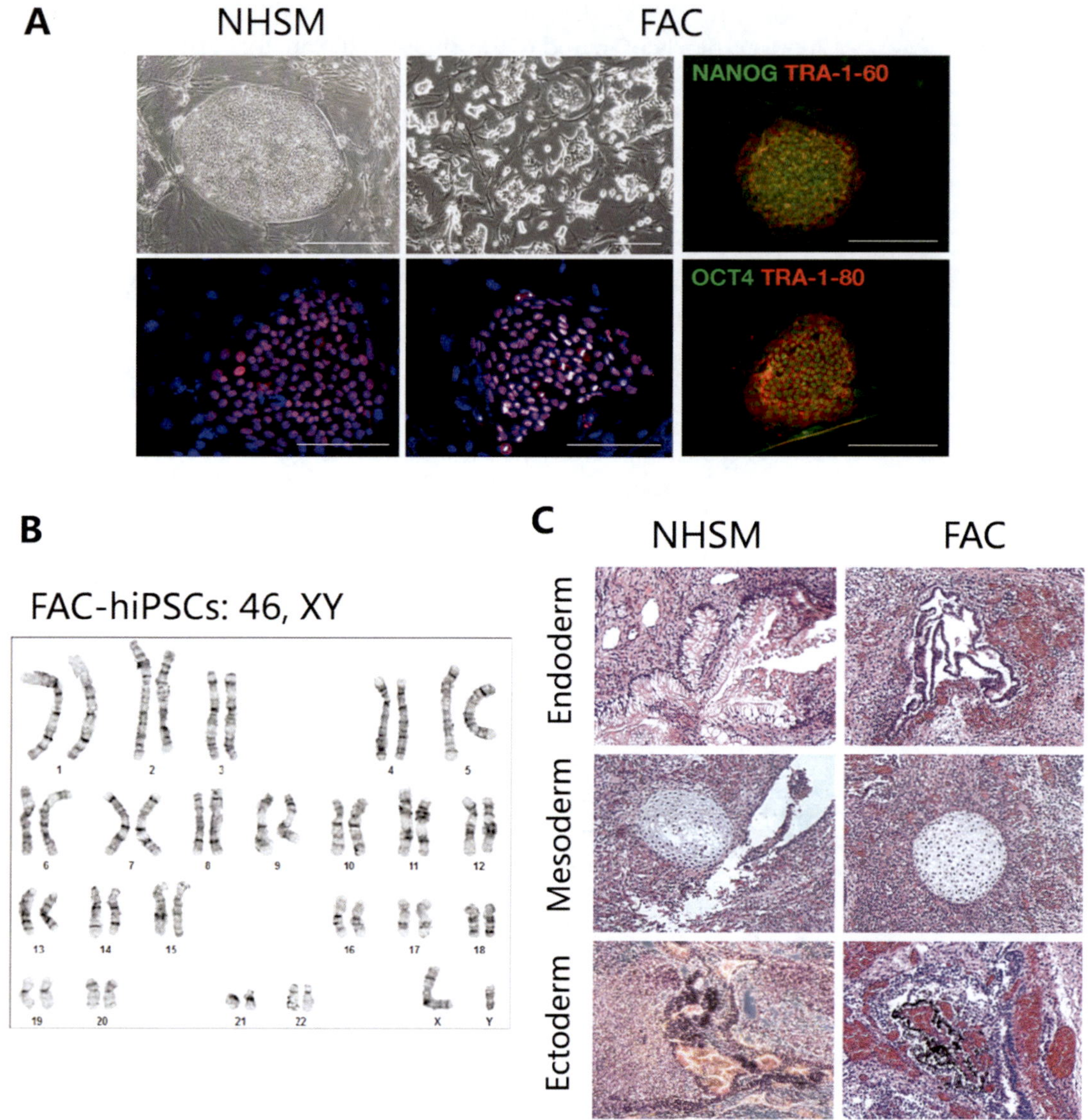

Fig. 2 Generation and characterization of NHSM- and FAC-hiPSCs. (**a**) Left panel: (Top) Representative bright-field images showing the colony morphology of naïve-like NHSM-hiPSCs. (Bottom) Representative immunofluorescence images of NHSM-hiPSCs stained with an anti-OCT4 antibody. Red: OCT4; blue: DAPI. Scale bar 100 μm. Middle panel: (Top) Representative bright-field images showing the colony morphology of intermediate FAC-hiPSC. (Bottom) Representative immunofluorescence images of FAC-hiPSCs stained with an anti-OCT4 antibody. Red: OCT4; blue: DAPI. Right panel: (Top) Representative immunofluorescence images of FAC-hiPSCs co-stained with anti-NANOG and TRA-1-60 antibody. Green: NANOG; red: TRA-1-60. (Bottom) Representative immunofluorescence images of FAC-hiPSCs co-stained with anti-OCT4 and TRA-1-80 antibody. Green: OCT4; red: TRA-1-80. Scale bar 100 μm. (**b**) Karyotype of intermediate FAC-hiPSC showing the normal set of chromosome (46, XY). (**c**) Representative images showing hematoxylin and eosin staining of histological sections derived from teratomas generated by NHSM-hiPSCs and FAC-hiPSCs. hiPSC-derived teratomas contained tissues from all three germ lineages: endoderm (top), mesoderm (middle), and ectoderm (bottom). Scale bar 100 μm

each. Mount the cells in ProLong Gold Antifade which contains DAPI.

7. Fluorescence imaging on inverted fluorescence microscope (Fig. 2a).

3.1.3 Karyotyping

1. Passage hiPSCs into a 100 mm dish. When hiPSCs reach to 50–60% confluence, feed the cell with fresh medium. The next day, change the medium containing 400 ng/mL demecolcine (Sigma), and incubate in cell incubator at 37 °C for 40 min–1 h (*see* **Note 6**).
2. Aspirate the medium and dissociate the cells into single cell with Accumax. Collect the cell into a 15 mL Falcon tube. Centrifuge at 200 × *g* for 5 min. Aspirate the medium, and wash the cell pellet with PBS once. Resuspend the cells with 8 mL pre-warmed hypotonic solution thoroughly and incubate at 37 °C water bath for 30 min.
3. Mix the cells with 1 mL pre-cooled fixative solution, and incubate at RT for 15 min. Centrifuge at 200 × *g* for 10 min at RT, aspirate the supernatant, and resuspend the cell pellet with 8 mL fixative solution. Incubate at RT for 30 min. Repeat fixation once. Incubate the samples in fixative solution at −20 °C overnight. (The samples can be stored for several months in fixative solution at −20 °C.)
4. Centrifuge at 200 × *g* for 10 min at RT and then aspirate the supernatant. Add moderate freshly prepared fixative solution (the volume of fixative solution depends on the amount of cell pellet and cell number) and resuspend the cell pellet by pipetting.
5. Prepare the microscope slides in advance. Clean the slide by 98% ethanol soak to remove oil and dirt, wash with dH_2O, and then ddH_2O. After drying, drop 3–4 drops of the cell onto pre-cleaned and cooled slides, spread the drops by blowing, and then dry the slides on fire immediately for several times or put the slides on the 60 °C hot plate for 1–2 min. Incubate the slides at 85 °C drying oven for 3–4 h or 60 °C drying oven overnight.
6. Prepare 0.005% (w/v) trypsin, dissolve in Earle's Balanced Salt Solution, transfer it to staining jar, and adjust the pH to 7.0–7.2 with 3% Tris. Incubate at 37 °C water bath for future use. Prepare 1× Giemsa stain solution.
7. Transfer the slides into prepared 0.005% trypsin solution and digest for 1–5 s (*see* **Note 7**). Put the slides in Earle's Balanced Salt Solution for few seconds and then rinse in water for few seconds to remove the trypsin.

8. Transfer the slides in 1× Giemsa solution for 10 min at RT. Wash the slide with water for 2–3 times to remove the remaining Giemsa. Dry the slide and check whether the G band is clear or not (*see* **Note 7**).
9. Analyze the metaphase spreads under a microscope. Find the metaphase at 20× objective of light microscope, and then change to 100× oil objective to count the chromosome number. Analyze the chromosome arrangement in each metaphase if capable (Fig. 2b).

3.1.4 Teratoma Assay

1. Pre-cool 1.5 mL Eppendorf tubes and 29G insulin syringe on ice (*see* **Note 8**).
2. Harvest 2×10^6 hiPSCs by Accumax dissociation, and centrifuge at $200 \times g$ for 5 min. Aspirate the medium, and then resuspend the pellet with 200 μL mixture of 100 μL culture medium and 100 μL Matrigel and placed on ice prior to injection.
3. Using pre-chilled syringe to inject the cells subcutaneously or intramuscularly into NOD SCID mice (*see* **Note 9**).
4. 8–12 Weeks later, harvest and dissect out the teratomas. Embed the teratomas in paraffin and process the teratomas for paraffin sectioning. Stain the teratomas paraffin sections with hematoxylin and eosin and determine the different tissue types under a microscope (Fig. 2c).

3.1.5 Generation of Fluorescently Labeled hiPSCs

1. Collect 2×10^6 dissociated hiPSCs and wash once with PBS. Here, we labeled NHSM naïve-like hiPSCs with hKO and FAC-hiPSC with GFP.
2. Resuspend the cell pellet with the Nucleofection/Plasmid mixture (P3 Primary Cell 4D-Nucleofector Kit, Lonza): 82 μL solution + 18 μL supplement1 + 2 μg plasmid pCAG-IP-humanized Kusabira Orange(hKO) or pEGIP.
3. Transfer the cells into cuvette and perform nucleofection using a 4D-Nucleofector (Program CB150, Lonza).
4. Seed the cells on puromycin-resistant DR4 MEFs. Change medium the next day.
5. Add 0.5–1 μg/mL puromycin to the culture medium 3 days post-transfection.
6. After 7–14 days' puromycin selection, manually pick drug-resistant and hKO- or GFP-positive colonies (*see* **Note 10**) and further expand as hKO- or GFP-labeled hiPSC lines.

3.2 Interspecies Human-Pig Chimera Formation

3.2.1 Generation of Pig Blastocysts

Generation of Pig Blastocysts Through Parthenogenesis

1. Collect prepubertal gilt ovaries at a local slaughter house. Rinse the ovaries with PBS-PVA (*see* **Notes 11** and **12**).
2. Aspirate the antral follicles (2–4 mm in diameters) using 20G needle attached to a 10 mL syringe.
3. Collect cumulus-oocyte complexes (COCs) and wash them in TCM-199 containing 0.1% PVA (*see* **Note 13**).
4. Incubate the COCs at 38.5 °C and 5% CO_2, 5% O_2, and 90% N_2 for 48 h in in vitro maturation medium (IVM: TCM-199-based medium) covered with mineral oil.
5. After IVM, strip the cumulus from maturated oocytes by incubation about 5 min in 1 mg/mL hyaluronidase dissolved with TL-HEPES-PVA and gentle pipetting (*see* **Note 14**).
6. Wash denuded oocytes with MEM containing 25 mM HEPES (*see* **Note 15**). Electrically activate the oocytes with two pulses of 120 V/mm for 40 μs, delivered by BTX Electro Cell Manipulator 2001 in a 0.5 mm chamber containing 0.3 M mannitol, 0.05 mM $CaCl_2$, 0.1 mM $MgSO_4$, and 0.1% (w/v) BSA (Fig. 3) (*see* **Note 16**).
7. After washing with PZM-5, incubate the oocytes in PZM-5 with 5 μg/mL cytochalasin B for 3 h to prevent the extrusion of second polar body (*see* **Note 17**) [18].
8. After activation, wash the embryos in culture medium and culture pig embryos in 500 μL of PZM-5 containing 0.3% BSA for 3–5 days at 38.5 °C in a humidified atmosphere of 5% CO_2, 5% O_2, and 90% N_2. After 4 days of culture, add 10% FBS into the culture medium (*see* **Note 18**).

Generation of Pig Blastocysts Through In Vivo Fertilization

1. Select the sows with a weaning-to-estrus interval of 3–4 days, and then induce superovulation by the intramuscular administration of 800 IU eCG 24 h after weaning (*see* **Note 19**).
2. Check estrus twice per day by exposing sows to a mature boar (nose-to-nose contact) and also apply manual back pressure. Females that exhibit a standing estrous reflex are considered to be in estrus.

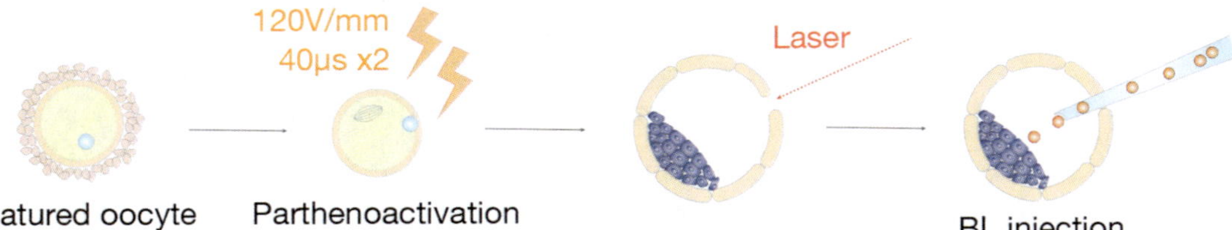

Fig. 3 Generation of human-pig chimeric embryos. Schematic of the experimental strategy for producing pig blastocyst obtained from parthenoactivation. Blastocysts were subsequently used for laser-assisted blastocyst injection of hiPSCs

3. At the onset of estrus, intramuscularly administer 750 IU of hCG to sows with clear signs of estrus at 48–72 h post-eCG administration.
4. Post-cervically inseminate the sows at 6 and 24 h after the onset of estrus. The insemination doses (1.5×10^9 spermatozoa in 45 mL) were prepared from sperm-rich fractions of the ejaculates extended in Beltsville thawing solution extender [19] and were stored for a maximum of 72 h at 18 °C (*see* **Note 20**).
5. Subject the sows to a midventral laparotomy on day 2 of the estrous cycle (day 0: onset of estrus). Sedate the donors with azaperone (2 mg/kg body weight, intramuscular) and anesthetize it with sodium thiopental (7 mg/kg body weight, intravenous), and then maintain with isoflurane (3.5–5%).
6. After exposure of the genital tract, count the corpora lutea on the ovaries and then flush each oviduct with 20 mL of protein-free embryo recovery medium consisting of TL-HEPES-PVA [20] with some modifications [21].
7. Collect embryos and wash them three times in TL-HEPES-PVA. Then, place the embryos in sterile EP tubes containing 1.5 mL of the same medium. Transport the embryos in a thermostatically controlled incubator at 38.5 °C immediately to the laboratory within 1 h after collection (*see* **Note 21**).
8. Evaluate the embryos for morphology under stereomicroscope at a magnification of 60×. In vivo-collected embryos may include zygotes with a single cell and two polar bodies, or 2–4-cell embryos, or morulae and blastocysts (*see* **Note 22**).
9. Transfer embryos to a Nunc 4-well multidish (40 zygotes per well) containing 500 μL of glucose-free NCSU-23 medium [22] supplemented with 0.3 mM pyruvate and 4.5 mM lactate for 24 h and then change to fresh NCSU-23 medium containing 5.5 mM glucose for an additional 5 days. Incubate the embryos at 38.5 °C, 5% CO_2, 5% O_2, and 90% N_2 in air and 95–97% relative humidity.
10. At day 5, supplement the embryo culture medium with 10% (v/v) FCS. Evaluate the in vitro embryo development under a stereomicroscope at 24 h and 6 days of culture to determine cleavage and blastocyst formation rates, respectively. When an embryo reaches the 2–4-cell stage it is considered as cleaved. An embryo with a well-defined blastocoel and an inner cell mass and trophoblast totally discernible is considered as blastocyst.

3.2.2 Microinjection of hiPSCs to Pig Morulae/Blastocysts and Embryo Culture

1. For morula injection, select the embryos with more than eight blastomeres and before compaction on days 3–4. For blastocyst injection, choose the embryos with an obvious blastocoel on days 5–6.

2. Choose hiPSCs showing an optimal undifferentiated morphology and proliferated exponentially (*see* **Note 23**). The cells should be at around 50–60% confluence. Dissociate hiPSCs from dishes, collect into a 15 mL Falcon tube, and centrifuge at 200 × *g* for 5 min. Remove the supernatant and wash the pellet with cell culture medium twice. Resuspend the cells with proper volume of cell culture medium (cell density approximately 2–6 × 10^5 cells/mL). The addition of Y-27632 Rock inhibitor is recommended. Place the cell suspension on ice before injection.
3. Add single-cell suspension to a 50 μL drop of cell culture medium containing the embryos to be injected. Then place it on an inverted microscope fitted with micromanipulators.
4. Collect individual cells into a blunt-end micropipette of 20–30 μm internal diameter connected to a manual hydraulic oil microinjector. Secure the embryo by a holding pipette (*see* **Note 24**).
5. Use a laser system to create a hole in the zona pellucida of the embryo. For blastocysts, apply another laser pulse to the trophectoderm in order to allow access to the blastocoele. Advance the micropipette containing the cells into the embryo and deposit ten cells in the blastocoel for blastocyst or perivitelline space for morulae (Fig. 3). Groups of 10–20 embryos can be manipulated simultaneously and each session is limited to 40 min (*see* **Note 25**).
6. Following cell injection, culture morulae and blastocysts in the respective cell culture medium for 4 h. Then, transfer the embryos to mix medium (1:1) of cell culture medium and PZM-5 containing 10% FBS for 20 h. After that, culture embryos in PZM-5 containing 10% FBS for another 24 h. Observe the ICM incorporation under a fluorescence microscope (Fig. 4).

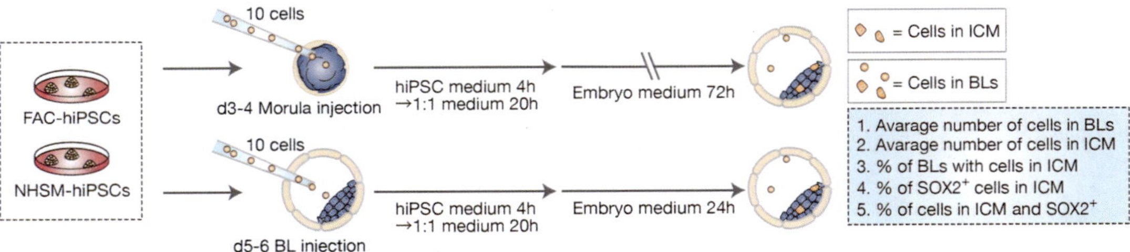

Fig. 4 Analysis of ICM contribution of hiPSC in preimplantation embryos. Schematic of the experimental procedures for morula or blastocyst injection of hiPSC. After injection, blastocysts were subsequently cultured in hiPSC medium for 4 h, and then transferred to medium composed of hiPSC medium and pig embryo culture medium (ratio 1:1) for 20 h. Finally, embryos were cultured in embryo medium until reaching blastocyst stage

7. For embryo transfer, the blastocysts after hiPSC injection are incubated in 500 μL of hiPSC cell culture medium for 3–4 h and then changed to mix medium (1:1 of NCSU medium and hiPSC medium) for an additional 20–22 h.

3.2.3 Pig Embryo Transfer

1. Use 6–7-month-old gilts as recipient pigs. Administer altrenogest orally for 15 days. 24 h after that, inject 800 IU eCG intramuscularly; 72 h later inject 750 U hCG intramuscularly to the sows. Check estrus as the same procedure described previously in Subheading "Generation of Pig Blastocysts Through In Vivo Fertilization." Females that exhibit estrus are used as recipients of surgical embryo transfer 5–6 days after hCG administration.
2. Prior to surgical procedure, fast the animals (food and water). Then, sows are subject to laparotomy. Induce the anesthesia by intramuscular administration of 2 mg/kg of Telazol prior to intubation. Maintain the sows at a surgical plane by inhalation of isoflurane (0.5–5% as needed to maintain anesthesia). Expose the ovary and uterus by a ventral medial laparotomy.
3. Load the injected blastocysts into a Gynetics embryo transfer catheter connected to a 1 mL syringe for transfer into the recipients. The embryo transfer medium is NCSU-23 supplemented with 10 mM HEPES, 0.4% (w/v) BSA, and 10%(v/v) FCS (*see* **Note 26**).
4. Load the embryo transfer catheter with air bubbles to separate 30 μL drop of medium that contains the embryo from two drops of medium before and after that embryo. All transfers are conducted in asynchronous (−24 h to embryo collection) recipients.
5. Transfer the embryos into the tip of a uterine horn (15–20 cm from the uterotubal junction) with the embryo transfer catheter inserted through the uterine wall, which is previously punctured with a blunt needle (*see* **Notes 27** and **28**).
6. Post-transfer, evaluate daily for the behavioral changes including signs of estrus beginning at 12 days post-transfer.

3.2.4 Collection of E21–28 Embryos by Pig Surrogate Cesarean Section

1. Diagnose the pregnancy by transabdominal ultrasonography on days 17–20 post-embryo transfer.
2. Deeply anesthetize the pregnant sows on days 23–25 post-transfer by intramuscular administration of 2 mg/kg of Telazol. And subsequently, euthanize them by intracardiac administration of 2.25 mL/kg euthanasia solution.
3. Make a midline longitudinal incision between the posterior pair of nipples. Locate the ovaries and uterus.

4. Occlude the cervix and the ovarian stalks with transfixing ligatures and the reproductive tract removed. Then place the genital tract in watertight plastic bags kept on ice. Transport the bags to the laboratory within 20 min.
5. Open the uterus and remove the fetuses from the placenta tissues. Number them in sequential orders (*see* **Note 29**).
6. Measure and weigh the fetuses and placenta individually.
7. Check the fluorescent emission of each fetus using an epifluorescence stereomicroscope.

3.3 Analysis of Chimeric Contribution of hiPSC to Pig Blastocysts

3.3.1 Immunostaining of Pig Blastocysts

1. The immunostaining is performed as previously described [23]. Collect the embryos reached to blastocyst stage. Wash the embryos with PBS containing 1 mg/mL PVA (PBS-PVA) three times.
2. Fix the embryo in 4% (w/v) PFA containing 1 mg/mL PVA for 15 min at RT.
3. After washing three times with PBS-PVA, permeabilize the blastocyst with PBS-PVA containing 1% Triton X-100 for 30 min.
4. Wash three times with PBS-PVA containing 0.1% Triton X-100 (washing buffer, WB). Block the embryos in PBS-PVA supplemented with 10% normal donkey serum for 1 h.
5. Incubate the embryos with primary antibody (SOX2 and HuNu) at 4 °C overnight.
6. Wash the embryo in washing buffer three times and incubate the embryos with fluorescent-conjugated secondary antibodies at RT for 1 h.
7. Mount the blastocyst on glass slide containing ProLong Gold Antifade solution and cover it with a coverslip. Image blastocysts with an inverted fluorescence microscope.

3.3.2 Evaluation of Chimeric Efficiency of hiPSCs to Pig Blastocysts

1. Calculate the percentage of blastocysts with human cells (hKO or HuNu positive) (Fig. 4).
2. Count the average number of human cells (HuNu positive) in each blastocyst (Fig. 4).
3. Count the average number of human cells (HuNu positive) locating in the inner cell mass (ICM) of each blastocyst (Fig. 4).
4. Calculate the percentage of SOX2-positive cell in human cells (HuNu) which locate in ICM (Fig. 4).
5. Calculate the percentage of human cells (HuNu positive) in the ICM in all SOX2-positive cell population (Fig. 4).

3.4 Analysis of Chimeric Contribution of hiPSC to Postimplantation Pig Embryos

3.4.1 Evaluation of Pig Embryo Developmental Status

1. Evaluate the morphology, embryo size, and weight of pig embryos injected with hiPSCs and compare them to non-injected control pig embryos (Fig. 5a).
2. Based on developmental status divide the embryos into normal and growth-retarded groups.

3.4.2 Fluorescence Imaging

1. Observe the GFP or hKO signal of the dissected fetuses under an epifluorescence stereomicroscope (Fig. 5b).

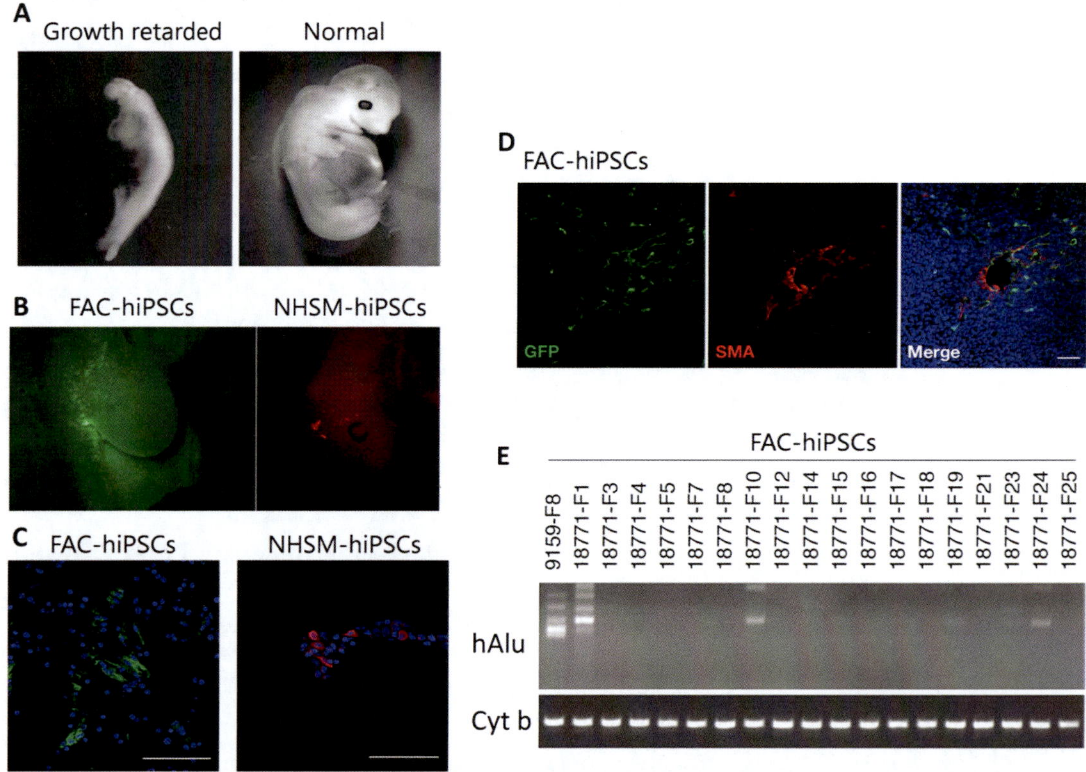

Fig. 5 Analysis of chimeric contribution of hiPSC in postimplantation embryos. (**a**) Representative bright-field images showing a normal-size day-28 pig embryo (right) and growth-retarded day-28 pig embryos (left). (**b**) Representative fluorescence images of GFP-labeled FAC-hiPSCs (left) and hKO-labeled NHSM-hiPSC (right) derivatives in normal-size day-28 pig embryo. (**c**) Representative immunofluorescence images showing the contribution of GFP-labeled FAC-hiPSCs (left) and hKO-labeled NHSM-hiPSCs (right) to normal-size day-28 pig embryos. Scale bar 100 μm. (**d**) Representative immunofluorescence images showing the chimeric contribution and differentiation of FAC-hiPSCs within a normal-size day-28 pig embryo. Embryo sections were stained with antibodies against GFP (green, left) and SMA (red middle). Right, merged images. Scale bar 100 μm. (**e**) Representative gel images showing genomic PCR analyses of pig embryos derived from blastocyst injection of FAC-hiPSCs (surrogates #18771) using a human-specific Alu primer. A pig-specific primer Cyt b was used for loading control

3.4.3 IHC Analysis of Fluorescence and Lineage Markers

1. Fix the dissected embryos in 4% PFA and incubate at 4 °C for 4 h for growth-retarded embryos and 1–2 days for the normal-sized embryos.
2. Place the embryos in 30% sucrose solution overnight until the embryos sink to the bottom of the tube. And then embed them in OCT compound. Freeze on dry ice and store in −80 °C.
3. Cut the embryo samples to make sections on a Leica cryostat: 10 μm thick for small-sized embryos and 20 μm for normal-sized embryos.
4. Perform the immunohistochemistry by using antigen retrieval solution (HistoVT one, Nacalai Tesque) according to the manufacturer's instructions. Briefly pour antigen retrieval solution in a glass container, and heat up to 70 °C using a water bath. Then incubate the dried frozen section in the heated antigen retrieval solution for 20 min at 70 °C. Wash frozen section with PBS three times followed by standard staining procedures (*see* **Note 30**).
5. Block the slides in PBST supplemented with 10% normal donkey serum for 1 h. Incubate primary antibody (fluorescent marker: Kusabira Orange2, GFP, or lineage marker: CK8, EpCAM, CAM: CK8, EpCAM) at 4 °C overnight.
6. Wash the slides in washing buffer three times and incubate the slides with fluorescent-conjugated secondary antibodies at RT for 1 h. Check the signal under a fluorescent microscope and capture immunofluorescence images using a confocal microscope (Fig. 5c, d).

3.4.4 Genomic PCR

1. Genomic PCR was carried out for the detection of human cells in pig fetuses.
2. Extract the genomic DNAs using DNeasy Blood and Tissue Kit (QIAGEN) according to the manufacturer's instructions (*see* **Note 31**).
3. Use PrimeSTAR GXL DNA polymerase (Takara) to perform the genomic PCRs (*see* Table 1).

   Add all the components above (Table 1) into 0.2 mL PCR tube. Then transfer the tube into PCR thermocycle for amplification.

   PCR conditions as follows:

   (a) 98 °C for 30s.

   (b) 98 °C for 5 s.

   (c) 55 °C for 15 s.

   (d) 68 °C for 10 s, repeat **steps 2–4** for 35 cycles.

   (e) 68 °C for 5 min.

   (f) 4 °C hold.

Table 1
Composition of PCR reaction mixture

Component	Volume	Final concentration
5× PrimeSTAR GXL buffer	5 μL	1×
dNTP mixture (2.5 mM each)	4 μL	200 μM each
Forward primer (Cyt b or hAlu-F)	0.5 μL	0.2 μM
Reverse primer (Cyt b or hAlu-R)	0.5 μL	0.2 μM
PrimeSTAR GXL Polymerase	2 μL	1.25 U/25 μL
Template (pig genomic DNA)	1 μL	50 ng
Sterilized distilled H_2O	12 μL	
Total	25 μL	

4. Load 10 μL PCR product on a 1% agarose gel.
5. Use a pig-specific primer Cyt b as loading control and set a group of negative control with no genomic DNA and positive control with human cells (*see* **Note 32**). Human-specific Alu sequence primers are used for the detection of human cells (Fig. 5e). Primer sequence is provided below:

   Human-specific Alu sequence primer:

   F: 5′-GGATTACAGGCGTGAGCCAC-3′.

   R: 5′-GATCGCGCCACTGCACTCC-3′.

   Pig-specific Cyt b mtDNA primer:

   F: 5′-GACCTCCCAGCCCCCTCAAACATCTCATCATGATGAAA-3′.

   R: 5′-GCTGATAGTAGATTTGTGATGACCGTA-3′.

4 Notes

1. The number of starting HFFs can affect the efficiency of episomal vector nucleofection and reprogramming.
2. The nucleofection kit and program can be further optimized for more efficient reprogramming. The percentage of GFP-positive cells can be used to evaluate the nucleofection efficiency.
3. Freshly prepared MEFs (prepared 1 day prior to use) are strongly recommended due to long term of the reprogramming process.

4. Choose the colonies which show more homogenous morphology. Make sure that the picked colonies can be further maintained after several rounds of passaging.
5. Even though NHSM and FAC culture condition allow single-cell passage of hiPSC, the cells are still sensitive to single-cell dissociation, especially at early passages. The addition of Rock kinase inhibitor Y-27632 greatly improves the single-cell cloning efficiency.
6. Choosing the time point with sufficient numbers of metaphase is very critical. Make sure that the cells are not too confluent and change the medium 1 day before demecolcine treatment.
7. Optimize the time of trypsin treatment. If the chromosomes are very stiff without obvious band, then increase the time for trypsin treatment. If the edge of the chromosome is rough and vague, try to decrease the time. After optimization, karyotype all the other samples.
8. Keep on ice for the whole process to avoid the gelation of Matrigel.
9. Check the status of NOD-SCID mouse after injection of hiPSCs. NOD-SCID mouse is immune deficient and is extremely sensitive to infection by a broad range of pathogenic and opportunistic microorganisms and special care is needed to take care of them and be sure to use aseptic handling techniques.
10. Perform a kill curve to determine the dosage used for selection. Also use lower dosage to maintain the stable transfectants because sometimes cells may lose the construct.
11. Add antibiotics to the ovary washing solution to avoid bacteria contamination.
12. Keep the ovaries in PBS-PVA in a water bath at 38.5 °C without changing to other medium and temperature.
13. Choose good COCs and oocytes with compact cumulus cell. Do not use the COCs with the darker and morphologically abnormal cumulus cells.
14. Accurately control the time of hyaluronidase treatment. If the time is too long, then it is not good for oocytes. However, if the time is too short, then the digestion will not be sufficient to remove all the cumulus cells. The remaining cumulus will not be good for embryo manipulation.
15. Wash the oocyte in drops and make sure that hyaluronidase is not remained due to its toxicity to embryos.
16. The parameter of electric stimulation for parthenoactivation can be optimized according to activation efficiency and embryo survival.

17. Check the pronucleus in embryos. Diploid parthenogenetic embryos always have two pronuclei which can be observed under inverted microscope.
18. The embryos should be thoroughly washed in medium at every medium-changing step to avoid remaining components from previous medium.
19. It is necessary to wean the sows to synchronize the estrus in donors. And also try to keep a weaning-to-estrus interval of 3–4 days, since the estrus is very important for efficient oocyte induction and fertilization.
20. High quality of sperms makes efficient fertilization and embryo development. Here we use in vivo fertilization due to high levels of polyspermic fertilization.
21. Oocytes and embryos are very sensitive to low temperature. So, it is important not to manipulate the embryos and oocytes outside for too long and keep the warm stage turning on at the temperature of 38.5 °C. Also, for the transportation of ovaries and embryos, use a thermally controlled container and carefully monitor the temperature.
22. The embryo harvested may be at a different stage; some are zygotes, and some may reach to 2–4-cell stage or morula. Keep all embryos for in vitro culture until morula or blastocyst stage for hiPSC injection.
23. The status of hiPSC is critical for its contribution in pig embryos. Also, the cell survival will be enhanced by the presence of ROCKi Y-27632 in culture medium during embryo injection and the first 4-h culture of embryos. Keep the cell suspension on ice before and during pig embryo injection.
24. Use blunt-end injection micropipette instead of sharp one. Also, the diameter of the injection pipette is critical for the embryo injection. If the diameter pipette is too large, big hole on zona pellucida is needed for injection, which may compromise embryo quality. Otherwise it is difficult to split cells from the injection pipette and also control the pressure and flow stream within injection pipette.
25. Do not keep the embryos outside for too long since they are sensitive to temperature. Divide the blastocyst into small groups and manipulate as quickly as possible.
26. Do not prepare the embryo in embryo transfer catheter too early to minimize the time outside.
27. The embryo transfer can be done unilaterally, because two sides of the uterine horns are connected.
28. The manipulation should be quick and gentle and reduce tissue damage and exposure of uterus as much as you can.

29. It is recommended to also keep the placentas, measure, weigh, and analyze their developmental status.
30. It is necessary to optimize the staining conditions for different tissues and antibody.
31. Homogenate the tissues before lysis and protein K treatment to make sure efficient DNA extraction if necessary. Add enough water to dissolve the DNA if it is sticky.
32. Negative and positive control is extremely important for successful detection of hiPSCs. Increase the UV exposure if the band is very weak due to limited chimerism.

Acknowledgment

We would like to thank Salk Waitt Advanced Biophotonic Core for technical advice on imaging analysis and Salk Stem Cell Core for providing cell culture reagents. We would like to thank May Schwarz and Peter Schwarz for administrative help. This work was supported by Universidad Católica San Antonio de Murcia (UCAM), the Larry L. Hillblom Foundation, Paul F. Glenn Foundation, and the Moxie Foundation.

References

1. Evans MJ, Kaufman MH (1981) Establishment in culture of pluripotential cells from mouse embryos. Nature 292(5819):154–156
2. Thomson JA et al (1998) Embryonic stem cell lines derived from human blastocysts. Science 282(5391):1145–1147
3. Takahashi K, Yamanaka S (2006) Induction of pluripotent stem cells from mouse embryonic and adult fibroblast cultures by defined factors. Cell 126(4):663–676
4. Takahashi K et al (2007) Induction of pluripotent stem cells from adult human fibroblasts by defined factors. Cell 131(5):861–872
5. Brons IG et al (2007) Derivation of pluripotent epiblast stem cells from mammalian embryos. Nature 448(7150):191–195
6. Tesar PJ et al (2007) New cell lines from mouse epiblast share defining features with human embryonic stem cells. Nature 448 (7150):196–199
7. Nichols J, Smith A (2009) Naive and primed pluripotent states. Cell Stem Cell 4 (6):487–492
8. Wu J et al (2015) An alternative pluripotent state confers interspecies chimaeric competency. Nature 521(7552):316–321
9. Wu J, Izpisua Belmonte JC (2016) Stem cells: a renaissance in human biology research. Cell 165(7):1572–1585
10. Ying QL et al (2008) The ground state of embryonic stem cell self-renewal. Nature 453 (7194):519–523
11. Huang Y et al (2012) In Vivo differentiation potential of epiblast stem cells revealed by chimeric embryo formation. Cell Rep 2 (6):1571–1578
12. Gafni O et al (2013) Derivation of novel human ground state naive pluripotent stem cells. Nature 504(7479):282–286
13. Takashima Y et al (2014) Resetting transcription factor control circuitry toward ground-state pluripotency in human. Cell 158 (6):1254–1269
14. Theunissen TW et al (2014) Systematic identification of culture conditions for induction and maintenance of naive human pluripotency. Cell Stem Cell 15(4):471–487
15. Wu J, Izpisua Belmonte JC (2015) Dynamic pluripotent stem cell states and their applications. Cell Stem Cell 17(5):509–525
16. Wu J et al (2017) Interspecies chimerism with mammalian pluripotent stem cells. Cell 168 (3):473–486.e15

17. Tsukiyama T, Ohinata Y (2014) A modified EpiSC culture condition containing a GSK3 inhibitor can support germline-competent pluripotency in mice. PLoS One 9(4):e95329
18. Yoshioka K, Noguchi M, Suzuki C (2012) Production of piglets from in vitro-produced embryos following non-surgical transfer. Anim Reprod Sci 131(1–2):23–29
19. Pursel VG, Johnson LA (1975) Freezing of boar spermatozoa: fertilizing capacity with concentrated semen and a new thawing procedure. J Anim Sci 40(1):99–102
20. Funahashi H, Ekwall H, Rodriguez-Martinez H (2000) Zona reaction in porcine oocytes fertilized in vivo and in vitro as seen with scanning electron microscopy. Biol Reprod 63 (5):1437–1442
21. Martinez EA et al (2014) Successful non-surgical deep uterine transfer of porcine morulae after 24 hour culture in a chemically defined medium. PLoS One 9(8):e104696
22. Petters RM, Wells KD (1993) Culture of pig embryos. J Reprod Fertil Suppl 48:61–73
23. Ross PJ et al (2008) Polycomb gene expression and histone H3 lysine 27 trimethylation changes during bovine preimplantation development. Reproduction 136(6):777–785

Chapter 9

Embryonic Chimeras with Human Pluripotent Stem Cells

Alejandro De Los Angeles, Masahiro Sakurai, and Jun Wu

Abstract

Human pluripotent stem (PS) cells can be isolated from preimplantation embryos or by reprogramming of somatic cells or germline progenitors. Human PS cells are considered the "holy grail" of regenerative medicine because they have the potential to form all cell types of the adult body. Because of their similarity to humans, nonhuman primate (NHP) PS cells are also important models for studying human biology and disease, as well as for developing therapeutic strategies and test bed for cell replacement therapy. This chapter describes adjusted methods for cultivation of PS cells from different primate species, including African green monkey, rhesus monkey, chimpanzee, and human. Supplementation of E8 medium and inhibitors of the Tankyrase and GSK3 kinases to various primate PS cell media reduce line-dependent predisposition for spontaneous differentiation in conventional PS cell cultures. We provide methods for basic characterization of primate PS cell lines, which include immunostaining for pluripotency markers such as OCT4 and TRA-1-60, as well as in vivo teratoma formation assay. We provide methods for generating alternative PS cells including region-selective primed PS cells, two different versions of naïve-like cells, and recently reported extended pluripotent stem (EPS) cells. These derivations are achieved by acclimation of conventional PS cells to target media, episomal reprogramming of somatic cells, or resetting conventional PS cells to a naïve-like state by overexpression of KLF2 and NANOG. We also provide methods for isolation of PS cells from human blastocysts. We describe how to generate interspecies primate-mouse chimeras at the blastocyst and postimplantation embryo stages. Systematic evaluation of the chimeric competency of human and primate PS cells will aid in efforts to overcome species barriers and achieve higher grade chimerism in postimplantation conceptuses that could enable organ-specific enrichment of human xenogeneic PS cell derivatives in large animals such as pigs and sheep.

Key words Pluripotent stem cells, Primed pluripotent stem cells, Embryonic stem cell, Induced pluripotent stem cell, Extended pluripotent stem cells, Nonhuman primates, Primates, Region-selective, Interspecies chimeras, FGF, WNT, GSK3, Tankyrase, TNKS1/2, Naïve-like pluripotent stem cells, LCDM, 5iLAF, t2iL, KLF2, NANOG, Reprogramming, Human pluripotent stem cells, Monkey pluripotent stem cells, Primate pluripotent stem cells, OCT4, SOX2, KLF4, LMYC, LIN28, p53

1 Introduction

Embryonic pluripotency spans a continuum of cellular states, ranging from a naïve state, which reflects the unrestricted developmental potential of the preimplantation epiblast, to a primed state, with characteristics of the differentiating postimplantation epiblast

Insoo Hyun and Alejandro De Los Angeles (eds.), *Chimera Research: Methods and Protocols*, Methods in Molecular Biology, vol. 2005, https://doi.org/10.1007/978-1-4939-9524-0_9, © Springer Science+Business Media, LLC, part of Springer Nature 2019

[1–3]. A rigorous test for evaluating stem cell potency is to introduce cultured PS cells into host embryos and assess their developmental potential [2]. Synchronization of the developmental stage of donor PS cells with recipient host embryos is critical for efficient chimera generation [4]. For example, mouse naïve embryonic stem (ES) cells produce chimeras when injected into morulae or blastocysts, whereas mouse epiblast stem (EpiS) cells engraft gastrula-stage embryos, but not vice versa [5–9]. As an alternative assay of stem cell potency, PS cells may be evaluated for their capacity to form interspecies chimeras. Rodent interspecies chimeras have been generated using naïve PS cells [10–14]. However, it remains unclear whether naïve PS cells from primate species can generate interspecies chimeras [13, 15–18].

The generation of human PS cells, which include ES cells from preimplantation human embryos as well as induced pluripotent stem (iPS) cells obtained through cellular reprogramming of somatic cells, has catapulted forward regenerative medicine [19–22]. Conventional human PS cells have been suggested to embody a primed pluripotent state because they share epigenetic and biological characteristics with mouse EpiS cells rather than mouse ES cells [8, 23]. Evidence suggests that conventional PS cells from nonhuman primates (NHP) also reside in a primed state. Like human PS cells, they share many defining features with mouse EpiS cells and most notably fail to form chimeras when injected into preimplantation embryos [24].

Recently, a number of studies have attempted to adjust the developmental stage of human PS cells by culturing them in media containing multiple kinase inhibitors. Such putative human naïve-like PS cells possess molecular characteristics associated with mouse ES cells [15, 17, 25–28]. It has been speculated that the naïve state of human pluripotency may exhibit an enhanced developmental potential when compared with conventional human PS cells. However, it remains unclear how to exploit the putatively enhanced developmental potential of human naïve pluripotency for regenerative medicine. Like rodent naïve PS cells, human naïve-like PS cells can potentially be used to generate interspecies chimeras, which may one day provide a means for producing functional human tissues via interspecies blastocyst complementation [29]. A requisite for successful interspecies blastocyst complementation with human PS cells is the ability of human PS cells to chimerize a developing embryo from another species. Initial results suggest that interspecies chimerism with a subset of naïve-like human PS cells is limited, which may reflect species barriers beyond matching developmental timing [13, 16, 17]. To realize this potential of naïve-like human PS cells, future studies are warranted to evaluate the influence of all possible culture parameters and to develop effective strategies to lower species barriers to enable efficient primate-nonprimate interspecies chimerism.

This chapter details the methods to derive and maintain PS cells from different primate species, including African green monkey, rhesus monkey, chimpanzee, and humans. Subtle differences between the PS cells of different primate species require refinement of culture parameters to confer experimental consistency. Here, we provide adjusted culture systems for conventional PS cells from different primate species by adding Tankyrase and GSK3 inhibitors and E8 medium to various PS cell media containing FGF2 [30, 31]. These adjustments reduce line-dependent tendencies for spontaneous differentiation.

Further, we provide protocols for generating alternative PS cell types with distinct molecular and functional features. These alternative PS cell types include region-selective PS cells (FGF2/R1 media), two different types of naïve-like PS cells (t2iL + DOX, 5iLAF media) [17, 28], and extended PS cells (LCDM medium) [18]. The generation of these cell types is achieved by acclimation of conventional PS cells to target media, episomal reprogramming of somatic cells, or resetting of conventional PS cells via ectopic expression of KLF2 and NANOG. Additionally, we describe how to derive human EPS cells from human blastocysts. Cultivating cells propagated in multi-inhibitor conditions stabilizes alternative primed, naïve-like, and extended states of pluripotency in human and NHP cells.

2 Materials

1. 293FT cells (ATCC).
2. Activin A.
3. Analgesic (e.g., 30 mg/kg ibuprofen; 0.05–0.1 mg/kg buprenorphine HCL).
4. Anesthetic (100 mg/kg ketamine and 10 mg/kg xylazine).
5. Anti-Nanog antibody (R&D Systems, AF1997).
6. Anti-Oct4 antibody (Santa Cruz, 8628).
7. Anti-SSEA4 antibody (DSHB, MC-813-70).
8. Anti-TRA-1-81 antibody (eBiosciences, 14-8883-82).
9. Antiseptic (e.g., Betadine solution).
10. B27 Supplement.
11. Beta-mercaptoethanol.
12. CF1 mouse embryonic fibroblasts (Thermo Fisher Scientific, A34180 or A34181).
13. CHIR99021.
14. Collagenase IV.
15. (S)-(+)-Dimethindene maleate.
16. Dispase.

17. DMEM/F12 medium.
18. Doxycycline hyclate (dox).
19. E8 medium (Thermo Fisher Scientific, A1517001).
20. 70% Ethanol.
21. FGF2.
22. Forceps, Watchmaker's #5, two pairs.
23. FUW-tetO-loxP-hKLF2 plasmid (Addgene, 60850).
24. FUW-M2rtTA (Addgene, #20342).
25. FUW-tetO-loxP-hNANOG plasmid (Addgene, 60849).
26. Cell-culture-grade gelatin.
27. Glutamax.
28. Human chorionic gonadotropin (hCG).
29. Hypodermic needle, 26 gauge 1/2 inch.
30. Hypodermic needle, 30 gauge 1/2 inch.
31. IM-12.
32. Inverted microscope with two micromanipulators.
33. IWR1.
34. KnockOut serum replacement (KSR) (Life Technologies, 10828).
35. L-Ascorbic acid 2-phosphate.
36. L-Glutamine.
37. 10 cm Petri dishes.
38. Leukemia inhibitory factor (LIF).
39. KSOMaa medium (Millipore, MR-121-D).
40. M2 medium (Millipore, MR-121-D) or other HEPES-buffered medium.
41. Matrigel.
42. Mercury.
43. Microcaps disposable micropipettes, Drummond 1-000-0500.
44. Microcaps disposable micropipettes, Drummond 1-000-1000.
45. Minocycline.
46. N2 Supplement (Thermo Fisher Scientific 17502048).
47. Nonessential amino acids (NEAA).
48. Neurobasal medium.
49. Mineral oil, embryo-tested light (Sigma, M8410).
50. P2 Primary Cell 4D-Nucleofector kit.
51. PD0325901.
52. 35mm Petri dishes.
53. pMD2.G (Addgene #12259).

54. pMDLg/pRRE (Addgene #12251).
55. Pregnant mares' serum gonadotropin (PMSG, e.g., from NIH National Hormone & Peptide Program; Prospec, HOR-272).
56. 12% Polyvinylpyrrolidone (Sigma, PVP360) dissolved with water (Sigma, W1503).
57. Polybrene.
58. pRSV-REV (Addgene #12253).
59. Scissors, fine.
60. Serrefine clamp.
61. StemFlex Medium Kit (Gibco, A3349401).
62. Stereomicroscope.
63. Suture, surgical silk (size 5-0).
64. Syringe, 1 mL.
65. TrypLE (Thermo Fisher Scientific 12605028).
66. Tungsten needle.
67. WH-4-023.
68. Wound clip and applier.
69. Y-27632.

2.1 KSR-CH/R1 Medium (Feeders) [31]

For 500 mL, combine 400 mL DMEM/F12 medium, 100 mL knockout serum replacement, 1% nonessential amino acids, 0.1 mM β-mercaptoethanol, 1% penicillin-streptomycin (optional), 20 ng/mL FGF2, 1 μM CHIR99021 (GSK3 inhibitor), 2.5 μM IWR1 (Tankyrase inhibitor).

2.2 mTESR1-CH/R1 Medium (Feeder Free) (Adapted from [31])

For 550 mL, combine 400 mL mTESR1 basal medium, 100 mL mTESR1 supplement, 50 mL E8 medium, 1 μM CHIR99021 (GSK3 inhibitor), 2.5 μM IWR1 (Tankyrase inhibitor).

2.3 CTFR Medium (mTESR1 Base) [9]

Prepare (mTESR1 base version) according to [32]: 20 ng/mL FGF2 and 2.5 μM IWR1.

2.4 Region-Selective Medium (N2B27 Base) [9]

For 500 mL, combine 240 mL DMEM/F12 medium, 240 mL neurobasal medium, 2.5 mL N2 supplement, 5 mL B27 supplement, 1% GlutaMAX, 1% nonessential amino acids, 0.1 mM β-mercaptoethanol, 1% penicillin-streptomycin, 5 mg/mL BSA (optional) or 5% knockout serum replacement (KSR, optional), 20 ng/mL FGF2, 2.5 μM IWR1.

2.5 t2iL + DOX Medium [17, 28]

For 500 mL, combine 240 mL DMEM/F12 medium, 240 mL neurobasal medium, 5 mL N2 supplement, 10 mL B27 supplement, 10 ng/mL human LIF, 1 μM CHIR99021, 1 μM PD0325901, 2 μg/mL DOX.

2.6 5iLAF Medium [17, 34]

For 500 mL, combine 240 mL DMEM/F12 medium, 240 mL neurobasal medium, 5 mL N2 supplement, 10 mL B27 supplement, 20 ng/mL human LIF, 20 ng/ml, 1 μM PD0325901, 1 μM IM-12, Enzo, 0.5 μM SB590885, 1 μM WH-4-023, 10 μM Y-27632, and 20 ng/mL activin A (0.5% KSR (GIBCO) can be included to enhance conversion efficiency. FGF2 (R&D systems, 8 ng/mL) enhances the generation of OCT4-ΔPE-GFP+ cells from the primed state, but it is dispensable for maintenance of naïve human ESCs.

2.7 LCDM Medium (iPSC Version) [18]

For 500 mL, combine 240 mL DMEM/F12 medium, 240 mL neurobasal medium, 5 mL N2 supplement, 10 mL B27 supplement, 10 ng/mL recombinant human LIF (L), 5 μM CHIR99021 (C), 2 μM (S)-(+)-dimethindene maleate (D), 2 μM minocycline hydrochloride (M), and 200 μg/mL L-ascorbic acid.

2.8 LCDM Medium (ESC Version) [18]

For 500 mL, combine 240 mL DMEM/F12 medium, 240 mL neurobasal medium, 5 mL N2 supplement, 10 mL B27 supplement, 10 ng/ml recombinant human LIF (L), 1 μM CHIR99021 (C), 2 μM (S)-(+)-dimethindene maleate (D), 2 μM minocycline hydrochloride (M), 2 μM Y27632, 0.5–1 μM IWR1, and 5 mg/mL BSA. We recommend including the BSA in the culture medium. Different Lot # of BSA will need to be tested for EPS cultures.

2.9 Embryo Dissection Medium

DMEM (Life Technologies, 11995-040), 10% fetal bovine serum (FBS, GEMINI Bio-Products, 100-106), penicillin-streptomycin (Life Technologies, 15140-122).

2.10 Embryo Culture Medium

Ham's F12 Nutrient Mix, GlutaMAX Supplement (Life Technologies, 31765-035), and N2 Supplement (Life Technologies, 17502-048) at a ratio of 100:1. Embryo culture medium is prepared by mixing rat serum (Harlan, B.4520) at a ratio of 1:1 with F12 (N2).

3 Methods

3.1 Maintenance of Primate PS cells

3.1.1 Conventional Culture and Routine Passaging of Primate PS Cells

Primate PS cells are normally passaged every 4–5 days (*See* **Note 1**). It is generally better to passage before confluence and never allow cultures to overgrow. 6-Well plates are recommended for routine human and NHP PS cell maintenance.

Feeder-Based Culture (KSR-CH/R1) [32] (Fig. 1A, Left, and 1B, Left)

1. Prepare iMEF-coated plates at least 1 day prior to passaging PS cells.

   Prepare MEF medium: DMEM with 10% FBS, L-glutamine and penicillin/streptomycin.

2. Coat plate with 0.1% gelatin and incubate at room temperature or 37 °C for at least 15 min.

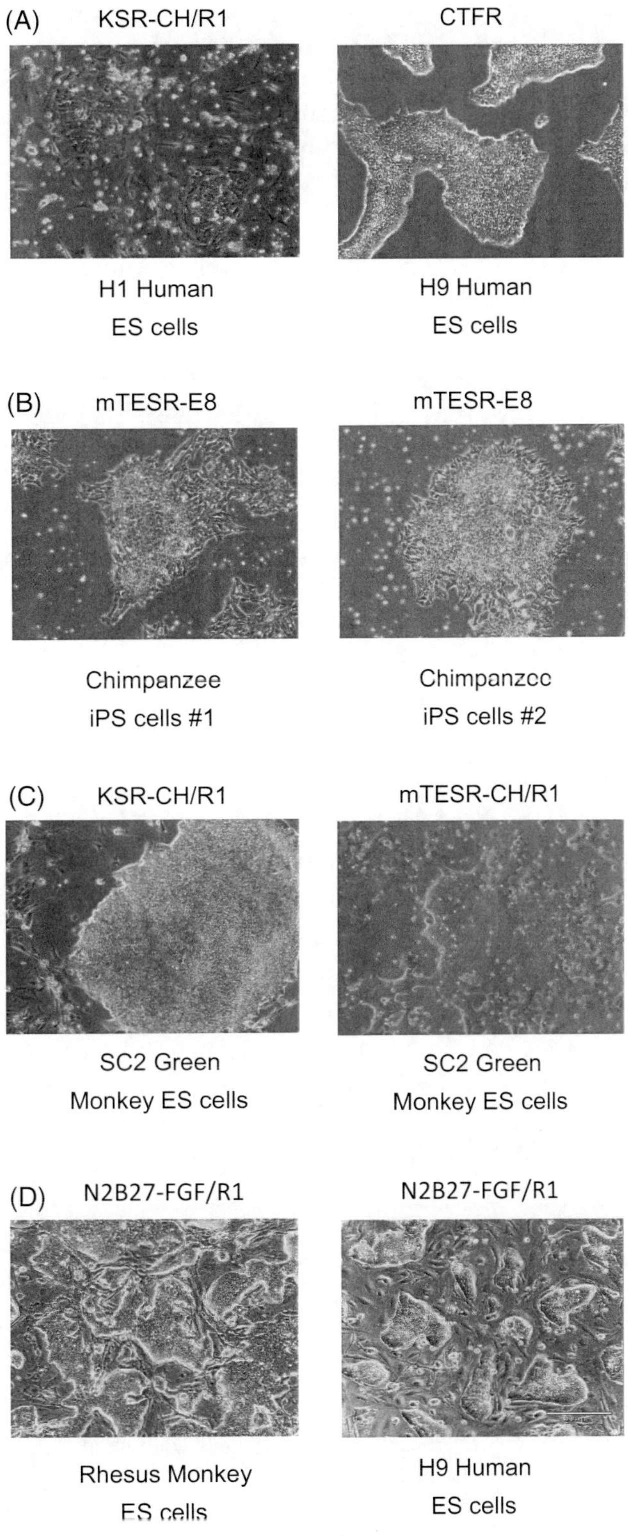

Fig. 1 Human and nonhuman primate pluripotent stem cells. (**A**) Human PS cells (left) H1 human ES cells cultured in KSR-CH/R1 medium; (right) H9 human ES cells cultured in CTFR medium. (**B**) Chimpanzee PS cells (left) chimpanzee iPS

3. Thaw iMEFs (commercial or prepared in-house) and seed onto gelatin-coated plates. Optimal feeder density may vary, but typically standard plating density for iMEFs that have been frozen and thawed before use is 0.90×10^5 cells/mL at 2.5 mL per well in a 6-well plate (or approximately 2.25×10^5 cells per well).
4. Incubate at 37 °C overnight to allow iMEFs to adhere.
5. Warm PS cell medium and PBS to room temperature before use.
6. Inspect PS cells before passaging and remove differentiated colonies as much as possible. Human and NHP PS cells are passaged enzymatically. Aspirate PS cell culture medium from each well and wash once with 2 mL PBS.
7. Add 1 mL of 1 mg/mL collagenase type IV and incubate at 37 °C for 5–10 min (split ratio 1:5).
8. Remove from 37 °C, and inspect visually; colonies should not have detached yet, but the edges should have. Aspirate collagenase solution.
9. Add 2 mL of culture medium per well.
10. Using a 1 mL pipette, scrape partially detached colonies.
11. Transfer cell suspension to a 15 mL conical tube. Centrifuge at $200 \times g$ for 3 min.
12. Aspirate the supernatant. Resuspend the cell pellet in PS cell medium.
13. Aspirate MEF medium from iMEF-coated plate. Try to remove as much MEF medium as possible.
14. Dispense human or NHP PS cell suspension into the wells according to the desired passage ratio.
15. Distribute PS cells evenly by moving the plate in several short, back-and-forth, and side-to-side motions. Incubate at 37 °C (5% CO_2).
16. Visually assess cultures daily to monitor growth until the next passage.

Fig. 1 (continued) cells cultured in mTESR-E8 medium; (right) chimpanzee iPS cells cultured in mTESR-E8 medium. (**C**) African green monkey PS cells; (left) SC2 green monkey ES cells cultured in KSR-CH/R1 medium; (right) SC2 green monkey ES cells cultured in mTESR-CH/R1 medium. (**D**) Rhesus monkey PS cells; (left) rhesus monkey ES cells cultured in N2B27 medium supplemented with FGF2 and IWR1 (*See* **Note 2**)

3.1.2 Feeder-Free Culture (mTESR1-CH/R1) (Ludwig et al., 2006; [30, 32]) (Fig. 1C, Right)

1. Prepare Matrigel-coated plates at least 1 h prior to plating PS cells.

   Note: Be sure to use cold/pre-chilled pipettes, tubes, and tips.
2. Add ~11 mL of chilled DMEM/F12 to a 15 mL conical tube.
3. Remove one 1.0 mg Matrigel aliquot from the freezer.
4. Using a P1000 pipette with a chilled 1000 μL tip, add 1 mL of cold DMEM/F12 to Matrigel aliquot.
5. Gently pipette to thaw and dissolve Matrigel.
6. Immediately transfer the DMEM/F12-Matrigel mixture to the 15 mL conical tube containing 11 mL cold DMEM/F12 medium.
7. Pipet up and down to mix well.
8. Plate 1 mL/well of a 6-well plate (total 12 mL should be sufficient for two 6-well plates).
9. Incubate for at least for 1 h at 37 °C.
10. Alternatively, plates can be stored at 4 °C for up to 2 weeks wrapped in parafilm. To use, place them in 37 °C incubator for at least 1 h before plating cells.
11. Warm PS cell medium and PBS to room temperature before use.
12. Inspect PS cell culture before passaging. Aspirate PS cell culture medium from each well.
13. Add 1 mL of dispase solution and incubate at 37 °C for 5–7 min (split ratio 1:5).
14. Remove from 37 °C, and inspect visually; colonies should not have detached yet, but the edges should have. Aspirate dispase solution.
15. Add 2 mL of culture medium per well.
16. Using a 1 mL pipette, scrape partially detached colonies.
17. Transfer cell suspension to a 15 mL conical tube. Centrifuge at 1000 rpm for 3 min.
18. Aspirate the supernatant. Resuspend the cell pellet in PS cell medium.
19. Aspirate Matrigel from new 6-well plate.
20. Dispense human or primate PS cell suspension into the wells according to the desired passage ratio.
21. Distribute PS cells evenly by moving the plate in several short, back-and-forth, and side-side motions. Incubate at 37 °C (5% CO_2).
22. Visually assess cultures daily to monitor growth until the next passage.

3.1.3 Cryopreservation

1. Collect PS cell colonies following the enzymatic passaging protocol. Gently break up colonies into small aggregates and spin down at 200 × *g* for 5 min.
2. Resuspend pellet in freezing medium. Place tubes in cryo-freezing container and store at −80 °C. The next day, transfer tubes in cryo-freezing container to a liquid nitrogen tank. Freezing medium is comprised of 10% DMSO, 40% PS cell medium, 50% FBS, and 10 μM Y-27632.

3.1.4 Thawing

1. Remove vial from liquid nitrogen storage and thaw in 37 °C water bath until about half of the vial is thawed.
2. Sterilize the outside of the tube with ethanol and bring it inside the cell culture hood. Use a P1000 pipette, in a dropwise manner, slowly add 500 μL–1 mL of pre-warmed PS medium to the vial, mix by gently pipetting up and down, and transfer into a 15 mL conical tube containing 8 mL of warm PS medium. Repeat this step several times until cells are completely thawed and transferred.
3. Spin down at 1000 rpm.
4. Resuspend in desired volume of medium containing 10 μM Y-27632, and gently add cell suspension into fresh 6-well plates seeded with iMEFs or coated with Matrigel.
5. Colonies could be ready to passage within 4–6 days.

3.2 Pluripotency Characterization

3.2.1 Immunostaining (OCT4, NANOG, SSEA4, TRA-1-81) (Adapted from [19])

1. Remove culture medium from the wells of a 6-well plate or 12-well plate.
2. Wash each well once with calcium/magnesium-containing PBS.
3. Aspirate PBS and add fixative (i.e., freshly prepared 4% paraformaldehyde in PBS) for 15 min at room temperature.
4. Conduct three 5-min washes with PBS.
5. Following final PBS wash, permeabilize by adding 0.1% Triton-X in PBS for 5 min. Next, add 10% FBS to 0.1% Triton-X/PBS. Add Triton-X/PBS/FBS mixture to wells for 1 h at room temperature to block.
6. Aspirate blocking solution and incubate with primary antibodies in 1% FBS and 0.1% Triton-X in PBS overnight. Primary antibodies: Nanog (R&D Systems, AF1997 at 1:20); Oct4 (Santa Cruz, 8628 at 1:100 concentration); SSEA4 (DSHB, MC-813-70 at 1:100); and TRA-1-81 (eBiosciences, 14-8883-82 at 1:100 concentration).
7. Next day, aspirate and wash three times with PBS for 5 min each.

8. Following final PBS wash, aspirate and incubate with fluorescent-labeled secondary antibodies at 1:500 dilutions for 1 h at room temperature.
9. Stain cells with DAPI.

3.2.2 Teratoma Assay

1. Expand PS cells sufficiently (number of required cells can be from as low as 100,000 to as high as 2,000,000).
2. Collect PS cells following an enzymatic passaging protocol.
3. Resuspend a desired number of cells in 40–50 μL of PBS.
4. Add an equal volume of ice-chilled Matrigel to the cell suspension and maintainon ice.
5. Using a 28.5 gauge syringe, inject the PS cell suspension into target anatomical locations (*See* **Note 3**). Anesthetize and handle animals according to the respective institutional animal use protocol. Carefully monitor animals as death from overdosing and hypothermia from long periods of anesthesia treatment can occur.
6. To enhance engraftment, allow the Matrigel/PBS cell mixture to solidify at the transplantation site before waking the animal. Keep animals under anesthesia on a 37 °C heat pad for 20 min postinjection.
7. After 8–12 weeks, examine the site of injection and recover teratoma from injection site.
8. Place teratoma in a 50 mL conical tube containing fixative (we typically use 4% PFA solution).
9. Transfer conical tube to a local pathology facility for analysis by an expert pathologist. Request hematoxylin and eosin (H&E) staining.

3.3 Conversion of Human and Primate PS Cells to an Alternative PS Cell State

3.3.1 Human and Primate Region-Selective PS Cells (Fig. 1A, Right; 1B, Right; 1C)

Human and primate rsPS cells are usually passaged every 5–7 days. For routine passaging, cryopreservation, and thawing of human and primate rsPS cells, follow methods as described in **Sections 1.0.2, 1.0.3, and 1.0.4.**, with the exception of using Accutase or TrypLE to conduct single-cell passaging. For the first few passages after conversion, we recommend adding 10 μM Y-27632 to the culture at least 1 h before and 24 h after each passage (*See* **Note 4**).

3.3.2 Generation of Naïve-Like hPS Cells

Generation of naïve-like cells by KLF2 and NANOG transduction of primed hPS cells (t2iL + DOX cells and 5iLAF cells) [17, 28] **(Fig. 2A, left)** (*See* **Note 5**).

1. Lentivirus production is prepared as follows: Transfect the following plasmids into 293FT cells: FUW-tetO-hNANOG,

FUW-tetO-KLF2, and packaging plasmids: pMDLg/pRRE (Addgene #12251), pRSV-REV (Addgene #12253), and pMD2.G (Addgene #12259) (separate 293FT wells for FUW-tetO-hNANOG and FUW-tetO-hKLF2).

2. 48 h After transfection, collect virus-containing supernatant from transfected 293FT cells and filter through a 0.45 μm filter.
3. Transduce primed PS cells with lentivirus in the presence of 10 μM Y-27632 and 6 μg/mL polybrene for 24 h.
4. After 24–48 h, transduced cells are replated onto CF1 MEF feeders. To generate t2iL + DOX cells, replate cells in 2iL medium containing 2 μg/mL DOX (Stemgent) (medium composition described below).
5. After 7–14 days, cultures are bulk passaged onto fresh MEF feeders, or manually picked onto fresh MEF feeders and expanded as t2iL + DOX hiPSCs (*See* **Note 6**).
6. To generate 5iLAF cells, replate t2iL + DOX hiPSCs onto fresh MEF feeders, fed with 5iLAF medium (medium composition described below), and DOX is withdrawn.

Generation of naïve-like and extended pluripotent stem (EPS) cells by direct reprogramming of human fibroblasts with episomal vectors and derivation from human blastocysts.

Generation of human extended pluripotent stem (EPS) cells by direct reprogramming of human fibroblasts with episomal vectors [18] (Fig. 2A, middle).

EPS-iPSCs are generated by reprogramming of human foreskin fibroblasts (HFF, ATCC, CRL 2429) with Yamanaka episomal vectors (OCT4, SOX2, KLF4, LMYC, LIN28, p53 shRNA) [33].

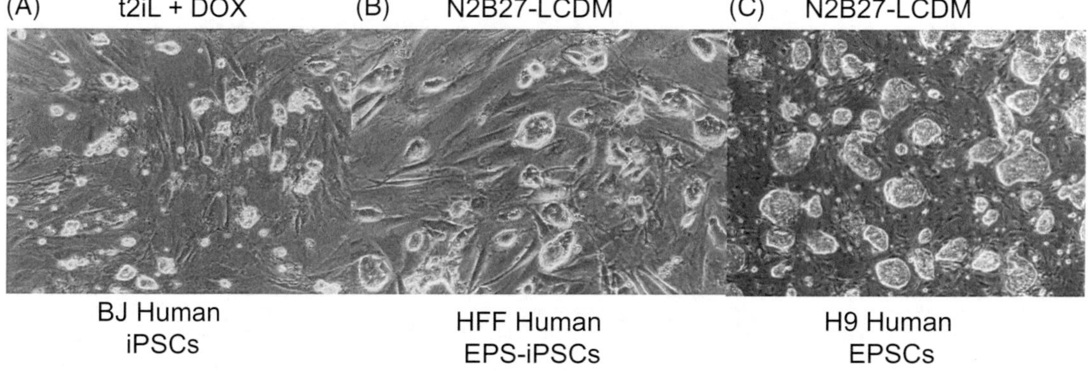

Fig. 2 Alternative human pluripotent stem cells. (**A**) Alternative human PS cells; (left) H1 human ES cells containing DOX-inducible KLF2 and NANOG transgenes maintained with t2iL (LIF, PD0325901, CHIR99021) + DOX; (middle) HFF human EPS-iPS cells cultivated in N2B27 basal medium supplemented with LCDM (LIF, CHIR99021, dimethindene maleate, minocycline); (right) H9 human ES cells cultured in N2B27 basal medium supplemented with LCDM

1. Prepare nucleofector solution + plasmids. Combine approximately 82 μL nucleofector solution with 18 μL supplement from Nucleofector kit. Add 1 μg of each Yamanaka episomal plasmid (3 μg DNA total) (episomal plasmids pCXLE-EGFP (Addgene, #27082), pCXLE-hOCT4-shp53 (Addgene, #27077), pCXLE-hSK (Addgene, #27078), pCXLE-hUL (Addgene, #27080)).
2. Collect 2×10^6 fibroblasts into a 15 mL conical tube and centrifuge at 1000 rpm for 5 min.
3. Aspirate supernatant carefully and try to leave behind the least volume possible.
4. Resuspend pellet in 100 μL nucleofector solution + Yamanaka plasmid suspension, and immediately add to a nucleofection cuvette.
5. Place cuvette in Amaxa Nucleofector II instrument and use program U-008.
6. Add 500 μL of warmed human fibroblast medium to the cuvette.
7. Add 1.5 mL human fibroblast medium to one well of a 6-well tissue culture plate.
8. Using the pipette provided in the Nucleofector kit box, transfer ~600 μL cell suspension from the cuvette to the well containing 1.5 mL of medium.
9. Place cells in a 37 °C incubator.
10. Five days post-nucleofection, replate HFFs onto mitotically inactivated MEFs.
11. The next day, change medium to conventional hPSC medium (medium composition described in earlier section) or EPS reprogramming medium comprised of KSR basal medium (Subheading 3.1) with the following adaptations: supplemented with 10% FBS, 10 ng/mL bFGF, 10 ng/mL LIF, 3 μM CHIR99021, and 10 μM Forskolin.
12. On day 12 post-transfection, either maintain cells in conventional hPSC medium or change EPS reprogramming medium to N2B27-LCDM medium ((iPSC version) composition described above).
13. Pick putative hiPSCs between days 20 and 30 and transfer to newly prepared MEFs for further cultivation. Monitor positive controls for the emergence of conventional human iPSCs in reprogramming plates maintained in conventional human PS cell medium. EPS-iPSCs should emerge around the same time in reprogramming plates maintained in N2B27-LCDM (iPSC version).

Conversion of Primed H9 hESCs into Human EPS Cells [18] (Fig. 2A, Right)

1. Digest primed H9 ES cells for conversion by removing conventional human PS cell medium from well of cells.
2. Wash with DMEM/F12. Aspirate DMEM/F12.
3. Add Accutase and incubate at 37 °C for 5–10 min.
4. Aspirate Accutase solution.
5. Wash again with DMEM/F12. Wash colonies off the dish surface by pipetting slowly.
6. Break large human primed PS cells into single cells.
7. Transfer cells into a 15 mL Falcon tube and centrifuge at 1000 rpm at room temperature for 3 min.
8. Aspirate supernatant.
9. Resuspend in an appropriate volume of conventional human PS cell medium.
10. Distribute cells onto plates with MEF feeders. Use conventional human PS cell medium supplemented with Y-27632 (10 μM).
11. 12 h After seeding, change medium from conventional human PS cell medium to N2B27-LCDM medium (for conversion of H9 ESCs, use 1 μM CHIR99021 instead of 5 μM CHIR99021).
12. Change N2B27-LCDM medium daily.
13. Within a few days, dome-shaped colonies should emerge.
14. 3–6 Days later, use Accutase to digest cells similarly as described above.
15. Centrifuge at 1000 rpm for 3 min.
16. Resuspend in an appropriate volume of N2B27-LCDM medium and seed into feeders.
17. After a few passages, cells that were cultured in LCDM medium should gradually proliferate well (*See* **Note 6**).

Human EPS Cell Derivation from Blastocysts [18]

1. For hatched blastocysts, use the following procedure:
2. Take plate containing iMEFs from the incubator.
3. Change the MEF medium into FBS-LCDM medium (knockout DMEM, 10% KSR, 10% FBS, 1% Glutamax, 1% NEAA supplemented with LCDM).
4. Plate whole embryos onto inactivated MEF feeder cells. Supplement FBS-LCDM medium with 10 μM Y-27632.
5. For unhatched blastocysts, use the following procedure:

6. Remove zona pellucida by protease.
7. Following disappearance of zona pellucida, transfer blastocysts into G2 PLUS medium.
8. Wash embryos 3–5 times to remove residual protease.
9. Seed embryo onto prepared MEF medium.
10. Two days later, change FBS-LCDM medium to N2B27-LCDM medium if embryo is attached to iMEFs. If embryo did not attach, remove half of FBS-LCDM medium and add a half volume of N2B27-LCDM medium ((ESC version) composition described above).
11. Outgrowths should become visible 4–7 days after being seeded onto iMEFs and should be dissociated mechanically into small clumps.
12. Seed cells onto fresh iMEF feeder cells supplemented with FBS-LCDM medium.
13. During the first few passages, it is recommended to dissociate nascent EPS cells mechanically and culture in FBS-LCDM medium supplemented with Y-27632. Transitioning to N2B27-LCDM (ESC version) medium can occur later.

3.3.3 Routine Passaging, Cryopreservation, and Thawing of Human and Primate Naïve-Like Cells

Human and primate naïve-like PS cells are normally passaged every 5–7 days. For routine passaging, cryopreservation, and thawing of human/primate naïve-like cells, follow methods as described above with the exception of using Accutase to conduct single-cell passaging. It is recommended to fastidiously monitor for karyotype changes (*See* **Notes** 7 and **8**).

3.4 Interspecies Chimera Generation

3.4.1 Grafting Human rsPS Cells to Postimplantation Epiblast

Isolation of Postimplantation Mouse E7.25-E7.5 Embryos

1. Timed-pregnant female mice (ICR) are prepared by mating with male mice. The next morning if vaginal plug is observed the embryos are aged E0.5.
2. Timed-pregnant female mice will be euthanized via CO_2 asphyxiation followed by cervical dislocation at E7.25–E7.5.
3. The abdomen will be sterilized with 70% ethanol followed by opening the abdominal cavity with fine scissors and forceps. The reproductive tract will be exposed.
4. The uterus can be removed by grabbing one of the uterine horns with a pair of forceps and cut the vaginal end of the uterus using a fine scissor. Next, remove the mesoterium and fat from the uterus.
5. Use a fine scissor to cut the connecting tissue between each embryo swelling and remove the muscle layer from each embryo swelling with a pair of fine forceps.
6. Transfer separated embryos to a new 10 cm Petri dish that has pre-warmed embryo dissection medium.

7. Pierce deciduum with a pair of fine forceps and open the forceps to push two sides apart. Use the tip of closed forceps to gently push out the embryos.
8. For epiblast grafting experimens no allantoic bud (OB) to early allantoic bud (EB) (OB-EB) stage embryos will be used.

Preparation of Mouse Embryos for Grafting

1. Use a pair of closed fine forceps and gently puncture the tip through the ectoplacental cone side of the yolk sac cavity and use it to affix the embryo to the bottom of the dish.
2. Insert the tip of another pair of closed fine forceps also into the same location within yolk sac and remove the Reichert's membrane, parietal endoderm, as well as trophoblast layer; leave the ectoplacental cone region intact (Fig. 3B, C).

Epiblast Grafting (*See* **Note 11–13**)

Human rsPS cells are grafted to OB- or OB-EB-stage epiblast with an aspirator tube assembly (Drummond) and a hand-pulled glass capillary (Drummond).

1. Human rsPS cells are washed twice with 1× PBS.
2. Under a stereoscope, use a 20 μL pipette tip to pick several colonies of human rsPS cells.
3. Use a tungsten needle to dissect each rsPS cell colony into small aggregates comprised of about 30–40 cells (Fig. 3A).
4. With a pair of fine forceps hold the embryo to be grafted, and insert the glass capillary containing rsPS cell clusters into anterior, posterior, or distal parts of the epiblast (Fig. 3C).
5. Make an opening in the epiblast region to be grafted by expelling out a small volume of dissection medium from the capillary tip.
6. Gently place a small clump of cells inside the opening.
7. Slowly pull out the glass capillary from the embryo.
8. Repeat **steps 4**–7 until all the embryos are grafted.
9. Continue cultivating the embryos in embryo culture medium [9].
10. Representative images of GFP-labeled H9 rsESCs after grafting and culturing for 36 h are shown in Fig. 4.

3.4.2 Blastocyst Chimerism of Naïve-Like PS Cells

Preparation of Embryo Donor Females

1. Induce superovulation to embryo donor females (C57BL/6, >8 weeks old) by intraperitoneal (ip) injection with 5 IU of PMSG, followed by the ip injection of 5 IU of hCG at 48 h post-PMSG.
2. Immediately after hCG injection, place one female in a cage with one stud male (F1 of C57BL/6 and DBA/1, >8 weeks old); males should be housed individually and each male should be mated with one female.

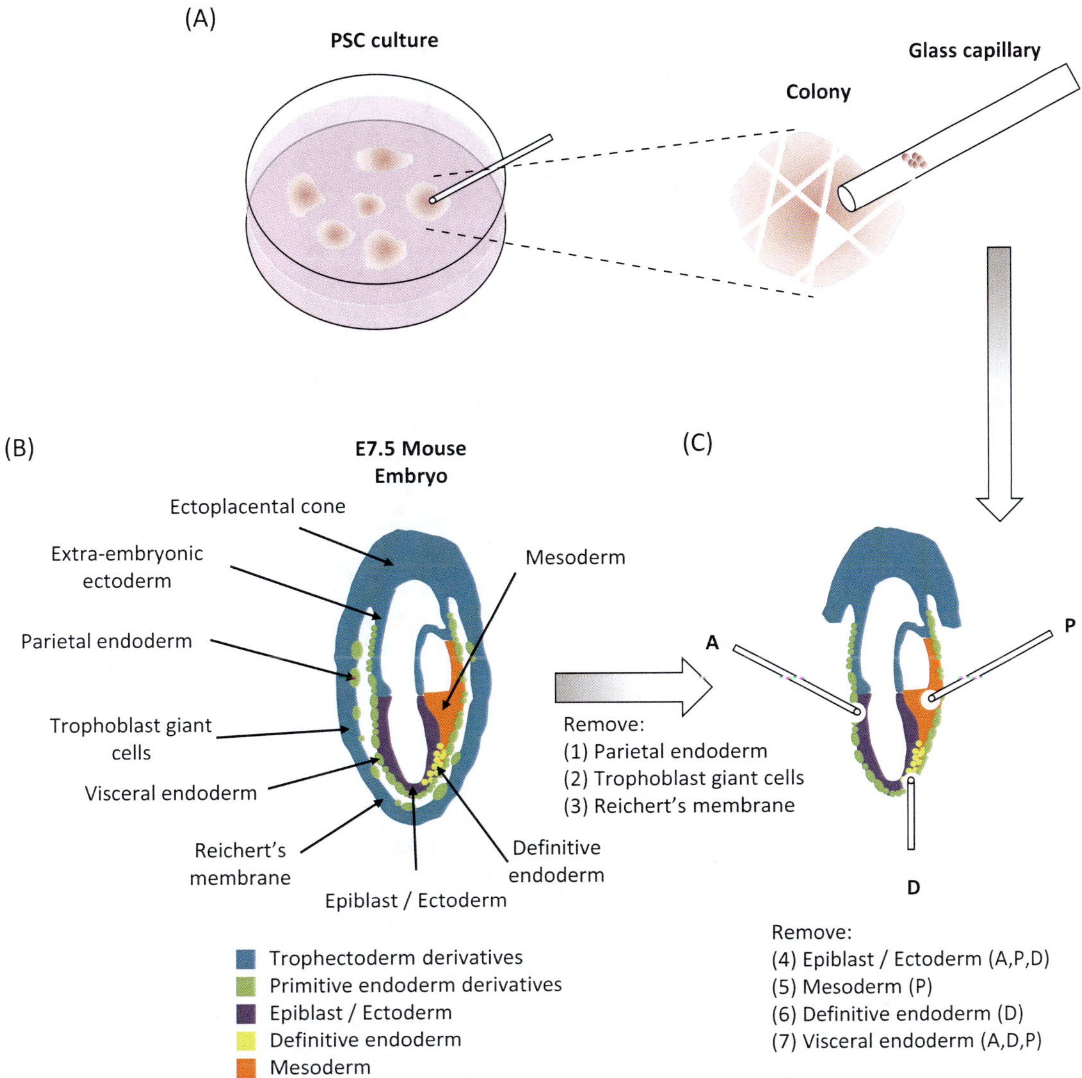

Fig. 3 Grafting rsPS cells to E7.5 mouse embryos. (**A**) Cut colonies of rsPS cells into small pieces with each aggregate containing about 30-40 cells before grafting. (**B**) A schematic drawing of an intact E7.5 mouse embryo. (**C**) A non-intact and nonviable E7.5 mouse embryo following dissection of parietal endoderm, trophoblast giant cells, and Reichert's membrane away from the embryo [1–3]. Additionally remove several sections to make openings for hosting embryos [4–7]. *A* Anterior, *P* Posterior, *D*, Distal

3. Next morning, check vaginal plug; the midday on the day of appearance of the plug is designated as 0.5 days postcoitum (0.5 dpc).
4. Separate females from males and house females together until embryo collection.

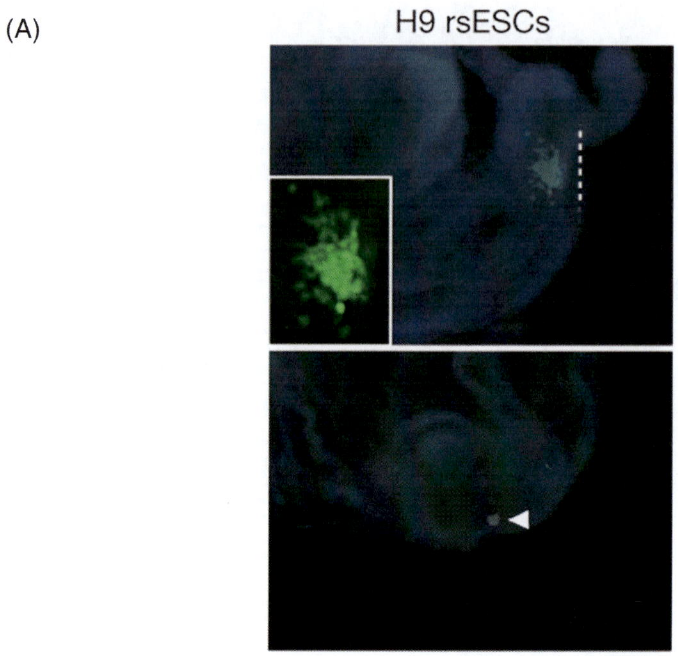

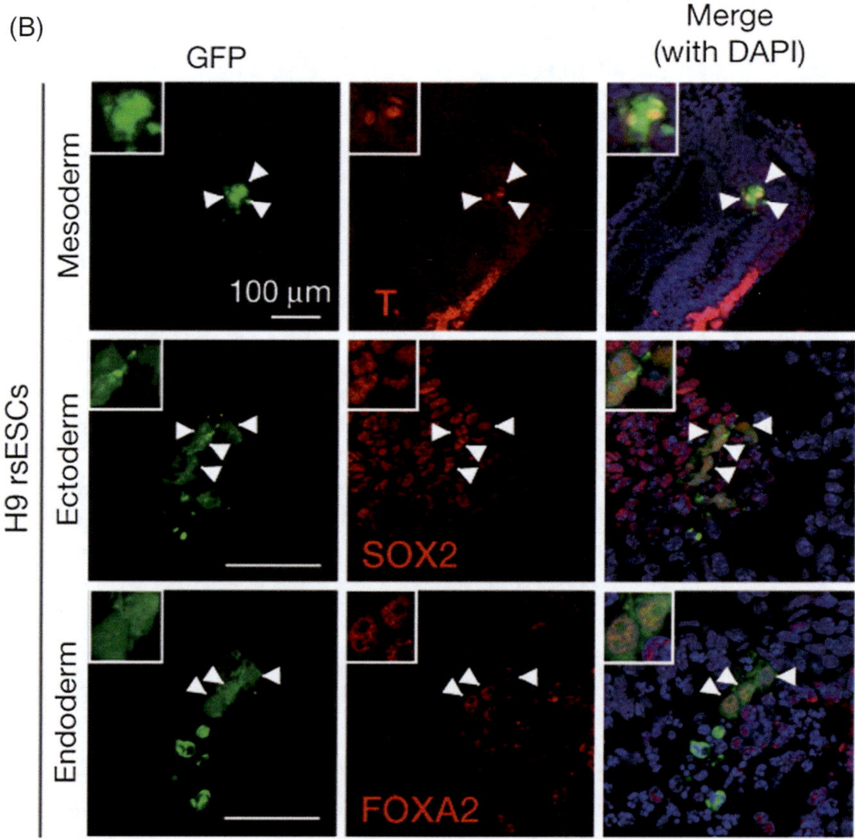

Fig. 4 Human region-selective pluripotent stem (rsPS) cells. (**A**) GFP-labeled H9 region-selective ES cells to the posterior (top) and anerior regions of mouse E7.5 epiblast. Arrowheads, a rsES cell clump. Dashed line,

Preparation of Surrogate Mother

If embryo donor females are found to have plugs, set natural mating for preparing pseudopregnant females as the surrogate mothers in the evening of the same day.

1. Set mating between ICR females as surrogate mothers (>8 weeks old) with vasectomized ICR males (>8 weeks old).
2. The next morning, check vaginal plug (0.5 dpc or E0.5).
3. Separate females from males and house females together.

Embryo Collection

Before dissection, prepare several droplets of KSOMaa medium (40–50 μL) covered with mineral oil in 35 mm Petri dishes. Medium droplets need to be equilibrated by placing the dishes in a humidified atmosphere of 5% CO_2 in the air at 37.0 °C overnight.

Perform handling of embryos with an aspirator tube assembly (Drummond) and a hand-pulled glass capillary (Drummond).

1. Euthanize donor females humanely at E2.75–3.0 by CO_2 and cervical dislocation.
2. Wipe abdomen with 70% ethanol.
3. Open the abdominal cavity and expose the reproductive tract.
4. Grasp the oviduct side of one of the uterine horns, cut between the ovary and the oviduct, and cut the uterus below the oviduct.
5. Place the collected oviducts into small droplets of M2 medium in a 35 mm dish at room temperature.
6. Under the stereomicroscope, hold the infundibulum side of the oviduct with one pair of fine dissection forceps and insert a blunt-end 30 gauge needle connected to 1 mL syringe containing M2 medium (*See* **Note 14**).
7. Flush out the oviducts with ~0.1 mL of M2 medium until embryos come out; embryos are in the 8-cell to morula stage.
8. Repeat **steps 6** and 7.
9. Place the 35 mm Petri dish containing the droplets of KSOMaa medium on the stage of stereomicroscope.
10. Select the embryos with good morphology using a pipette and transfer them into a droplet of KSOMaa medium.
11. Wash them through successive transfer between several droplets of KSOMaa medium.

Fig. 4 (continued) dispersed cells. Blue, DAPI [images reproduced with permission from *Nature*]. (**B**) Representative immunofluorescent images of posteriorly grafted GFP-labeled H9 rsES cells. Blue, DAPI. Insets, high magnification views. Arrowheads indicate T (brachyury)-, SOX2-, or FOXA2-positive derivatives of grafted cells [images reproduced with permission from *Nature*] (*See* **Note 9**)

12. After wash, transfer the embryos into droplets of KSOMaa medium for culture (ten embryos per each droplet).
13. Culture embryos in a humidified atmosphere of 5% CO_2 in the air at 37.0 °C overnight.

Microinjection of PS Cells

Holding and injection pipettes are made with a micropipette puller (Sutter Instrument) and a microforge (Narishige) or ordered from commercial sources (e.g., Origio and VitroLife). Here, the injection procedure with Piezo Impact Drive (Prime Tech) is described. On the day of injection, the PS cells should be prepared about an hour before the embryos are ready to be injected.

Before microinjection, prepare several droplets of M2 medium and 12% PVP medium (both 40–50 μL) covered with mineral oil in the lid of a 10 cm Petri dish (an injection chamber: Fig. 5A) (*See* **Note 15**).

Load a small quantity of mercury (around 1 cm long) into the injection pipette from the end of the pipette with a 26 gauge needle or a flexible plastic capillary tube connected to a 1 mL syringe.

Attach the holding and injection pipettes to the oil-filled injectors and expel the air from the tips of pipettes in the PVP medium droplet (Fig. 5Ba).

Wash and coat tips of the holding and injection pipettes with PVP medium until mercury and PVP medium move smoothly (Fig. 5Bb, c).

1. The next morning after embryo collection (E3.5), check blastocyst development. Examples of optimal and suboptimal blastocysts are shown in "Manipulating the Mouse Embryo: A Laboratory Manual (Third Edition)."
2. Wash the cells twice with 1× PBS.
3. Trypsinize the cells and resuspend them in 50–100 μL of KSOMaa or ES cell culture medium.
4. Place a few hundred single cells and expanded blastocysts into a 40 μL droplet of M2 on the lid of a 10 cm Petri dish.
5. Place the dish onto an inverted microscope fitted with micromanipulators.
6. Load about 100 cells into an injection pipette (15 μm inner diameter) connected to the microinjector (Fig. 5Ca).
7. Secure the blastocysts by a holding pipette with an ICM positioned at 9 o'clock (Fig. 5Ca).
8. Touch the end of the injection pipette gently to the surface of the zona pellucida (Fig. 5Ca).
9. Create a hole in the zona pellucida and trophectoderm with several times of piezo pulses (Fig. 5Ca).

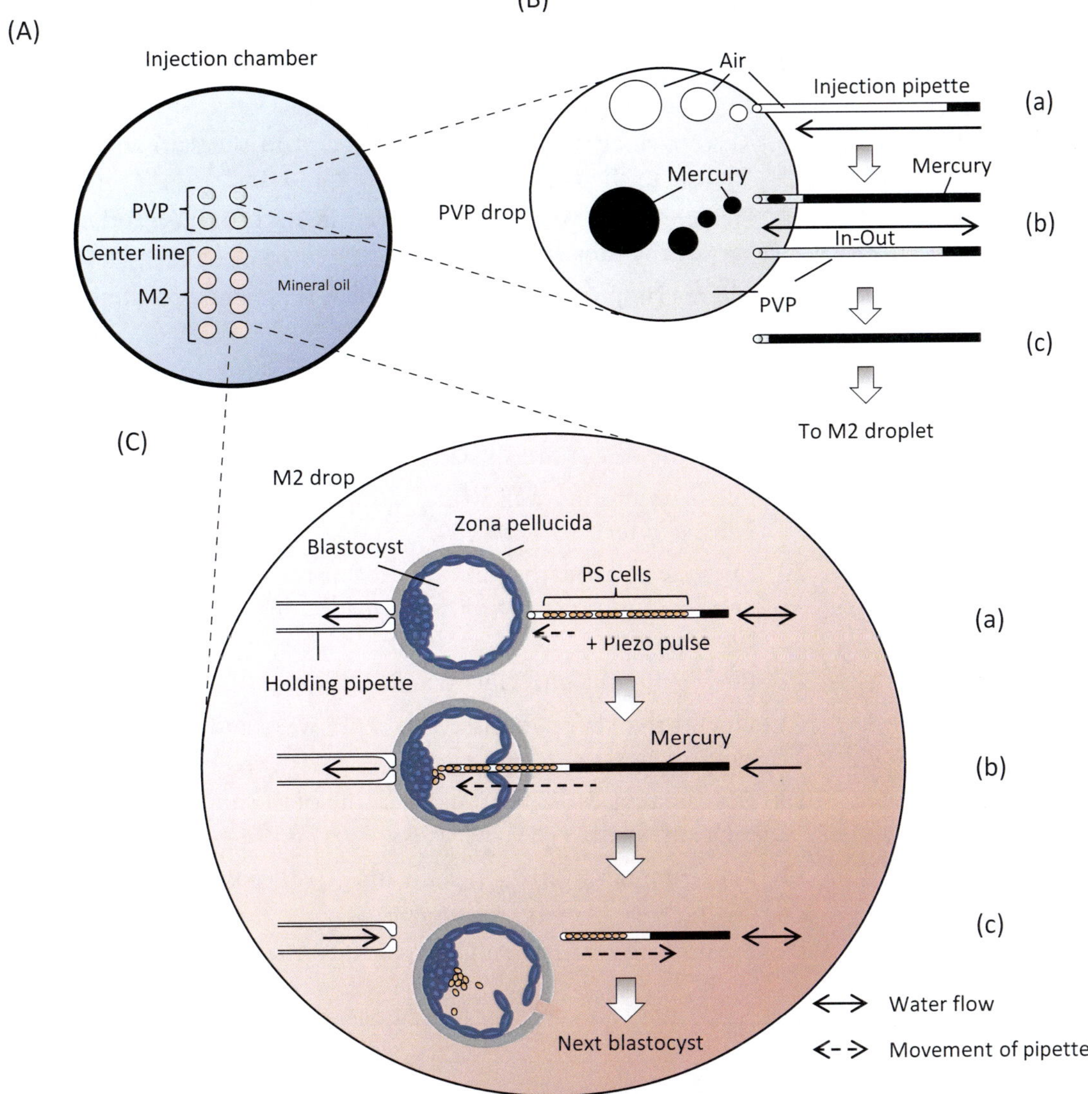

Fig. 5 Microinjection of PS cells. (**A**) An illustration of an injection chamber. (**B**) Preparation of injection pipette. In a droplet of 12% PVP medium, expel the air from the tip completely (**a**), and wash several times until mercury and PVP medium move smoothly (**b** and **c**). (**C**) Microinjection of PS cells to E3.5 embryos. In the droplet of M2 medium, load PS cells into the injection pipette, secure blastocyst by a holding pipette, and create a hole in zona pellucida and trophectoderm with piezo pulses (**a**). Insert the injection pipette and introduce 10–15 PS cells (**b**). Withdraw the injection pipette and release the embryo (**c**)

10. Insert the injection pipette and introduce 15–20 single cells into the blastocoel near the ICM (Fig. 5Cb).
11. Withdraw the injection pipette slowly and repeat **steps 7–10** for each blastocyst (Fig. 5Cc).
12. After microinjection, culture blastocysts in KSOMaa medium for at least 1 h before the embryo transfer (*See* **Note 16**).

Embryo Transfer (*See* **Notes 10, 16–21**)

1. Weigh and anesthetize a surrogate female at E2.5 and place the mouse on the lid of a 10 cm Petri dish.
2. Wipe the back of the surrogate female with 70% ethanol and apply antiseptic.
3. Make a small dorsal incision (<1 cm) in the skin and subsequently in the body wall just over the ovary.
4. Pick up the fat pad, pull out the ovary and oviduct, and expose part of the uterus horn.
5. Clip a clamp onto the fat pad which is attached to the ovarian bursa.
6. Place the mouse on the stage of stereomicroscope.
7. Load injected blastocysts into the pipette with air bubble.
8. Hold the top of uterus with dissection forceps.
9. Puncture uterine wall with a 26 gauge needle connected to a 1 mL syringe.
10. Keep an eye on the hole made by the needle, pull out the needle, and insert the pipette into the hole punctured toward the cervix.
11. Blow gently to introduce 15–20 blastocysts per surrogate.
12. Unclip the clamp, suture the body wall, and close skin with wound clips.
13. Repeat **steps 3–12** for the other side of the uterus horn and for each surrogate.
14. Keep animals warm on a heating pad postoperatively until recovery.
15. Treat postoperative pain with analgesic for 3–5 days.

4 Notes

1. For primed culture of human and primate PS cells, variation across species and cell lines has been observed with regard to stability and predisposition to differentiate. Therefore, it is important to adjust culture conditions appropriately for optimal propagation of each cell line.
2. Although not required for propagation, we observed that the addition of GSK3 and Tankyrase inhibitors, or Tankyrase inhibitor alone, to conventional human and primate PS cell culture media reduces line-dependent predisposition for spontaneous differentiation.
3. Note on the selection of anatomical locations for teratoma assay for testing pluripotency: sites are intra-sub-renal capsule, intramuscular, and subcutaneous.

4. For primed, naïve-like, and extended PS cell culture, it is critical to routinely karyotype to avoid expansion of karyotypically abnormal cells.
5. For generation of naïve-like and extended PS cell types, we observed substantial line-dependent variation when attempting to acclimate conventional human and primate PS cell to target medium. This is why we provide protocols for reprogramming factor-based derivation of naïve-like PS cells or EPS cells or resetting factor-based derivation of naïve-like PS cells (t2iL + DOX, 5iLAF). In contrast, we did not observe any variation for transitioning of conventional PS cells to region-selective PSC media.
6. It may be beneficial to pick undifferentiated colonies during primed to naïve-like or EPS cell conversions and expand in target medium. Additionally, it is beneficial to use the H9 hES cell line to gain experience with the conversion to different conditions.
7. 5iLAF cells and T2iL + DOX cells, while exhibiting molecular characteristics closest to human preimplantation embryos, have a reported predisposition toward karyotypic abnormalities and imprint loss. It is advised to monitor the karyotypes of these cell types [34].
8. Further optimizations of LCDM medium may be necessary, as LCDM cells in human and NHPs show some predisposition to differentiate. To minimize spontaneous differentiation, (1) avoid plating LCDM cells too sparsely, (2) use the proper quantity of freshly prepared MEF feeder cells, and (3) do not allow cells to overgrow [18].
9. When comparing different human PS cell types for interspecies chimera formation, it is essential to use cell lines from the same genetic background as well as multiple cell lines for each condition.
10. It is likely that evolutionary distance between species of donor PS cells and species of host embryo influences the degree of interspecies chimerism [9].
11. We have not tested all naïve-like conditions in primate cells. The methods and data regarding generation of naïve-like cells and interspecies chimeras have been tested predominantly in human species. It is expected that human PS cell experiments will translate to NHP PS cells but additional adjustments may be necessary. For example, vitamin C was added to NHSM condition for culture of cynomolgus monkey PS cells [35]. It should also be noted that interspecies chimera formation for rhesus monkey iPSCs propagated in a modified NHSM condition has been reported [36].

12. After epiblast grafting, FGF2/ACTIVIN A-cultivated mouse EpiS cells proliferate well in the distal and posterior region and to a lesser extent in the anterior portion. In contrast, although rsEpiS cells efficiently integrate in the posterior epiblast, they do not integrate well into the anterior and distal sections. Following an additional 36 hours of ex vivo culture, rsEpiS cells grafted to the posterior portion of the epiblast can proliferate and differentiate into endoderm, mesoderm, and ectoderm lineages.
13. Similarly to rodent rsEpiS cells, primate rsES cells can also proliferate and appropriately commence differentiation toward all three embryonic germ layers – ectoderm, mesoderm, and endoderm – after engraftment to the posterior portion of mouse E7.5 epiblasts.
14. For embryo collections, the opening of the oviduct (infundibulum) is swollen and can be located within the oviduct coil. Turn the collected oviduct around the M2 medium with fine forceps and slide the opening onto the flushing needle.
15. For microinjection of PS cells, an injection chamber should be kept at room temperature. Lower temperatures shrink blastocysts, which makes it harder to create holes. Higher temperatures cause PS cells to become sticky, which makes it harder to collect single cells. Minimize the damage to the blastocysts by aiming to insert the injection pipette at the junction between the trophoblast cells and by not touching the ICM with the pipette. If the PS cells start to stick to one other and to the tip of injection needle, it is advisable to be cautious with piezo pulses for making holes, because it becomes difficult to leave a certain number of cells inside the embryos. In this case, expel all cells and rinse needle again in the PVP medium. Expelling a small quantity of mercury and/or rinsing with piezo pulses could help remove sticky aggregates and clean the injection pipette. If injection needle gets too dirty, attach a new pipette. The blastocysts will collapse once the pipette is removed, which will facilitate contact between the introduced cells and the ICM. The blastocysts will re-expand after 1–3 h.
16. Although detailed methods for evaluating interspecies chimeric competency in mouse conceptus are not provided here, practitioners may interrogate chimerism in the mouse conceptus after embryo transfer and retrieval from murine foster mothers. We describe caveats below:
17. For embryo transfer, load the embryos into the pipette near the tip and close to each other in a minimal amount of media with air bubbles. Air bubbles introduced inside the uterus horn indicate successful embryo transfer. If there is no flow when blowing, check the tip of the pipette under the microscope. If it

is clogged, expel the embryos and reload into the same or a new pipette. The pups are usually born around E19.5 according to the schedule of surrogate mothers.

18. Growth retardation and embryonic lethality are frequently observed after retrieval of implanted interspecies chimeric embryos from foster mothers [13].

19. Currently available methods do not give rise to high-grade interspecies chimeras. When low-grade chimerism is observed following retrieval of postimplantation embryos, it is important to be cautious evaluating chimerism using the presence of fluorescently labeled donor cells. It is essential to exclude artifacts such as autofluorescence and uptake of markers by dead cells. It may be beneficial to use quantitative PCR analysis of human mitochondrial DNA [13, 16]. However, the presence of positive signal does not necessarily indicate functional integration of donor cells as DNA can be derived from dead cells or contamination during sample preparations [13, 16].

20. The presence of fluorescent labeled donor cells is not definitive proof for chimeric contribution as donor PS cell-derived cells often appear as a cluster and may not fully integrate into animal tissues. It is strongly recommended to confirm the function of donor cell derivatives whenever possible [13, 16].

21. Whenever possible, following introduction of naïve-like PS cells and EPS cells into preimplantation embryos, it is useful to analyze early postimplantation embryos (E5.5–E6.5) to assess colonization of the epiblast as well as the extraembryonic ectoderm for assessment of trophoblastic potential.

References

1. Nichols J, Smith A (2009) Naïve and primed pluripotent states. Cell Stem Cell 4:487–492
2. De Los Angeles A, Ferrari F, Xi R, Fujiwara Y, Benvenisty N, Deng H, Hochedlinger K, Jaenisch R, Lee S, Leitch HG, Lensch MW, Lujan E, Pei D, Rossant J, Wernig M, Park PJ, Daley GQ (2015) Hallmarks of pluripotency. Nature 525:469–478
3. Wu J, Izpisua Belmonte JC (2015) Dynamic pluripotent stem cell states and their applications. Cell Stem Cell 17:509–525
4. Cohen MA, Markoulaki S, Jaenisch R (2018) Matched developmental timing of donor cells with the host is crucial for chimera formation. Stem Cell Rep 10:1445–1452
5. Brons IG, Smthers LE, Trotter MW, Rugg-Gunn P, Sun B, Chuva de Sousa Lopes SM, Howlett SK, Clarkson A, Ahrlund-Richter L, Pedersen RA, Vallier L (2007) Derivation of pluripotent epiblast stem cells from mammalian embryos. Nature 448:191–195
6. Huang Y, Osorno R, Tsakiridis A, Wilson V (2012) In Vivo differentiation potential of epiblast stem cells revealed by chimeric embryo formation. Cell Rep 2:1571–1578
7. Kojima Y, Kaufman-Francis K, Studdert JB, Steiner KA, Power MD, Loebel DAF, Jones V, Hor A, de Alencastro G, Logan GJ et al (2014) The transcriptional and functional properties of mouse epiblast stem cells resemble the anterior primitive streak. Cell Stem Cell 14:107–120
8. Tesar PJ, Chenoweth JG, Brook FA, Davies TJ, Evans EP, Mack DL, Gardner RL, McKay RD (2007) New cell lines from mouse epiblast share defining features with human embryonic stem cells. Nature 448:196–199
9. Wu J, Okamura D, Li M, Suzuki K, Luo C, Ma L, He Y, Li Z, Benner C, Tamura I, Krause

MN, Nery JR, Du T, Zhang Z, Hishida T, Takahashi Y, Aizawa E, Kim NY, Lajara J, Guillen P, Campistol JM, Esteban CR, Ross PJ, Saghatelian A, Ren B, Ecker JR, Izpisua Belmonte JC (2015) An alternative pluripotent state confers interspecies chimaeric competency. Nature 521:316–321

10. Honda A, Choijookhuu N, Izu H, Kawano Y, Inokuchi M, Honsho K, Lee A-R, Nabekura H, Ohta H, Tsukiyama T et al (2017) Flexible adaptation of male germ cells from female iPSCs of endangered *Tokudaia osimensis*. Sci Adv 3:e1602179
11. Kobayashi T, Yamaguchi T, Hamanaka S, Kato-Itoh M, Yamazaki Y, Ibata M, Sato H, Lee YS, Usui J, Knisely AS, Hirabayashi M, Nakauchi H (2010) Generation of rat pancreas in mouse by interspecific blastocyst injection of pluripotent stem cells. Cell 142:787–799
12. Lee S-G, Mikhalchenko AE, Yim SH, Lobanov AV, Park J-K, Choi K-H, Bronson RT, Lee C-K, Park TJ, Gladyshev VN (2017) Naked mole rat induced pluripotent stem cells and their contribution to interspecific chimera. Stem Cell Rep 9:1706–1720
13. Wu J, Platero-Luengo A, Sakurai M, Sugawara A, Gil MA, Yamauchi T, Suzuki K, Bogliotti YS, Cuello C, Morales Valencia M, Okumura D, Luo J, Vilarino M, Parrilla I, Soto DA, Martinez CA, Hishida T, Sanchez-Bautista S, Martinez-Martinez ML, Wang H, Nohalez A, Aizawa E, Martinez-Redondo P, Ocampo A, Reddy P, Roca J, Maga EA, Esteban CR, Berggren WT, Nunez Delicado E, Lajara J, Guillen I, Guillen P, Campistol JM, Martinez EA, Ross PJ, Izpisua Belmonte JC (2017) Interspecies chimerism with mammalian pluripotent stem cells. Cell 168:473–486
14. Xiang AP, Mao FF, Li W-Q, Park D, Ma B-F, Wang T, Vallender TW, Vallender EJ, Zhang L, Lee J et al (2008) Extensive contribution of embryonic stem cells to the development of an evolutionarily divergent host. Hum Mol Genet 17:27–37
15. Gafni O, Weinberger L, Mansour AA, Manor YS, Chomsky E, Ben-Yosef D, Kalma Y, Viukov S, Maza I, Zviran A, Rais Y, Shipony Z, Mukamel Z, Krupalnik V, Zerbib M, Geula S, Caspi I, Schneir D, Shwartz T, Gilad S, Amann-Zalcenstein D, Benjamin S, Amit I, Tanay A, Massarwa R, Novershtern N, Hanna JH (2013) Derivation of novel human ground state naïve pluripotent stem cells. Nature 504:282–286
16. Theunissen TW, Friedli M, He Y, Planet E, O'Neil RC, Markoulaki S, Pontis J, Wang H, Iouranova A, Imbeault M, Duc J, Cohen MA, Wert KJ, Castanon R, Zhang Z, Huang Y, Nery JR, Drotar J, Lungjangwa T, Trono D, Ecker JR, Jaenisch R (2016) Molecular criteria for defining the naïve human pluripotent state. Cell Stem Cell 19:502–515
17. Theunissen TW, Powell BE, Wang H, Mitalipova M, Faddah DA, Reddy J, Fan ZP, Maetzel D, Ganz K, Shi L, Lungjangwa T, Imsoonthornruksa S, Stelzer Y, Rangarajan S, D'Alessio A, Zhang J, Gao Q, Dawlaty MM, Young RA, Gray NS, Jaenisch R (2014) Systematic identification of culture conditions for induction and maintenance of naïve human pluripotency. Cell Stem Cell 15:471–487
18. Yang Y, Liu B, Xu J, Wang J, Wu J, Shi C, Xu Y, Dong J, Wang C, Lai W, Zhu J, Xiong L, Zhu D, Li X, Yang W, Yamauchi T, Sugawara A, Li Z, Sun F, Li X, Li C, He A, Du Y, Wang T, Zhao C, Li H, Chi X, Zhang H, Liu Y, Li C, Duo S, Yin M, Shen H, Belmonte JCI, Deng H (2017) Derivation of pluripotent stem cells with in vivo and extraembryonic potency. Cell 169:243–257
19. Thomson JA, Itskovitz-Eldor J, Shapiro SS, Waknitz MA, Swiergiel JJ, Marshall VS, Jones JM (1998) Embryonic stem cell lines derived from human blastocysts. Science 282:1145–1147
20. Takahashi K, Tanabe K, Ohnuki M, Narita M, Ichisaka T, Tomoda K, Yamanaka S (2007) Induction of pluripotent stem cells from adult human fibroblasts by defined factors. Cell 131:861–872
21. Takahashi K, Yamanaka S (2006) Induction of pluripotent stem cells from mouse embryonic and adult fibroblast cultures by defined factors. Cell 126:663–676
22. Yu J, Vodyani MA, Smuga-Otto K, Antosiewicz-Bourget J, Frane JL, Tian S, Nie J, Jonsdottir GA, Ruotti V, Stewart R, Sluvkin II, Thomson JA (2007) Induced pluripotent stem cell lines derived from human somatic cells. Science 318:1917–1920
23. Hanna J, Cheng AW, Saha K, Kim J, Lengner CJ, Soldner F, Cassady JP, Muffat J, Carey BW, Jaenisch R (2010) Human embryonic stem cells with biological and epigenetic characteristics similar to those of mouse ESCs. PNAS 107:9222–9227
24. Tachibana M, Sparman M, Ramsey C, Ma H, Lee HS, Penedo MC, Mitalipov S (2012) Generation of chimeric rhesus monkeys. Cell 148:285–295
25. Chan YS, Goke J, Ng JH, Lu X, Gonzalez KA, Tan CP, Tng WQ, Hong ZZ, Lim YS, Ng HH (2013) Induction of a human pluripotent state with distinct regulatory circuitry that resembles preimplantation epiblast. Cell Stem Cell 13:663–675

26. Ware CB, Nelson AM, Mecham B, Hesson J, Zhou W, Jonlin EC, Jimenez-Caliani AJ, Deng X, Cavanaugh C, Cook S, Tesar PJ, Okada J, Margaretha L, Sperber H, Choi M, Blau CA, Treuting PM, Hawkins RD, Cirulli V, Ruohola-Baker H (2014) Derivation of naïve human embryonic stem cells. PNAS 111:4484–4489
27. Valamehr B, Robinson M, Abujarour R, Rezner B, Vranceanu F, Le T, Medcalf A, Lee TT, Fitch M, Robbins D, Flynn P (2014) Platform for induction of maintenance of transgene-free hiPSCs resembling ground state pluripotent stem cells. Stem Cell Rep 2:366–381
28. Takashima Y, Guo G, Loos R, Nichols J, Ficz G, Krueger F, Oxley D, Santos F, Clarke J, Mansfield W, Reik W, Bertone P, Smith A (2014) Resetting transcription factor control circuitry toward ground-state pluripotency in human. Cell 158:1254–1269
29. Wu J, Greely HT, Jaenisch R, Nakauchi H, Rossant J, Belmonte JC (2016) Stem cells and interspecies chimaeras. Nature 540:51–59
30. Chen G, Gulbranson DR, Hou Z, Bolin JM, Ruotti V, Probasco MD, Smuga-Otto K, Howden SE, Diol NR, Propson NE, Wagner R, Lee GO, Antosiewicz-Bourget J, Teng JM, Thomson JA (2011) Chemically defined conditions for human iPSC derivation and culture. Nat Methods 8:424–429
31. Ludwig TE, Bergendahl V, Levenstein ME, Yu J, Probasco MD, Thomson, JA (2006) Feeder-independent culture of human embryonic stem cells. Nat Methods 3:637–646
32. Kim H, Wu J, Ye S, Tai CI, Zhou X, Yan H, Li P, Pera M, Ying QL (2013) Modulation of beta-catenin function maintains mouse epiblast stem cell and human embryonic stem cell self-renewal. Nat Commun 4:2403
33. Okita K, Matsumara Y, Sato Y, Okada A, Morizane A, Okamoto S, Hong H, Nakagawa M, Tanabe K, Tezuka K, Shibata T, Kunisada T, Takahashi M, Takahashi J, Saji H, Yamanaka S (2011) A more efficient method to generate integration-free human iPS cells. Nat Methods 8:409–412
34. Pastor WA, Chen D, Liu W, Kim R, Sahakyan A, Lukianchikov A, Plath K, Jacobsen SE, Clark AT (2016) Naïve human pluripotent stem cells feature a methylation landscape devoid of blastocyst or germline memory. Cell Stem Cell 18:323–329
35. Chen Y, Niu Y, Li Y, Ai Z, Kang Y, Shi H, Xiang Z, Yang Z, Tan T, Si W, Li W, Xia X, Zhou Q, Ji W, Li T (2015) Generation of cynomolgus monkey chimeric fetuses using embryonic stem cells. Cell Stem Cell 17:116–124
36. Fang R, Liu K, Zhao Y, Li H, Zhu D, Du Y, Xiang C, Li X, Liu H, Miao Z, Zhang X, Shi Y, Yang W, Xu J, Deng H (2014) Generation of naïve induced pluripotent stem cells from rhesus monkey fibroblasts. Cell Stem Cell 15:488–497

Chapter 10

Neural Stem Cell Transplantation into a Mouse Model of Stroke

Alejandro De Los Angeles

Abstract

Stroke is the fifth leading cause of death among Americans each year. Current standard-of-care treatment for stroke deploys intravenous tissue-type plasminogen activator (tPA), mechanical thrombolysis, or delivery of fibrinolytics. Although these therapies have reduced stroke-induced damage, therapeutic options still remain limited. Transplantation of patient-specific neural stem (NS) cells represents a promising strategy for the treatment of stroke. Basic science research has shown that transplanted NS cells can differentiate in the brain of rodent models of stroke and promote behavioral recovery. Clinical trials exploring the feasibility of stem cell treatment for stroke are currently being conducted. However, questions remain regarding the optimal means of delivering NS cells, including cell dose, infusion speed, timing of transplantation, anatomic site, and imaging-assisted monitoring and guidance. Of the different available delivery modalities, intravascular NS delivery after stroke represents one practical approach. In this chapter, I provide methods for intravascular delivery of NS cells in a mouse model of stroke. The techniques involved include cell culture of NS cells, flow cytometry of NS cells, modeling stroke via unilateral common carotid artery occlusion, intra-arterial injection of NS cells into the brain, behavior analyses, and immunohistochemistry. Intra-arterial NS cell therapy has the potential to improve functional recovery after ischemic stroke.

Key words Stroke, VLA4, CD49d, CCR2, VCAM1, CCL2, Postnatal chimera, Neural stem cell, Common carotid artery, Hypoxia-ischemia, Mouse stroke model, C17.2 neural progenitors

1 Introduction

Stroke is the fifth leading cause of death among Americans each year [1]. The current treatment options for stroke involve fibrinolytics, mechanical thrombolysis, and intravenous tissue-type plasminogen activator (tPA). Despite the successful implementation of these treatments for routine treatment of stroke, additional therapeutics are needed to improve functional recovery after stroke. Thus, a strong impetus compels the biomedical community to develop new therapeutic modalities such as stem cell-based therapies [2]. Various studies provide proof-of-concept evidence for the feasibility of treating stroke with stem cells. Transplantation of NS cells into rodent models of stroke shows that NS cells differentiate

Insoo Hyun and Alejandro De Los Angeles (eds.), *Chimera Research: Methods and Protocols*, Methods in Molecular Biology, vol. 2005, https://doi.org/10.1007/978-1-4939-9524-0_10, © Springer Science+Business Media, LLC, part of Springer Nature 2019

in vivo and that transplanted mice show enhanced behavioral recovery when compared with non-transplanted mice [3, 4]. One day, transplantation of NS cells into patients after stroke may restore neurological function in patients.

A central question for developing cell therapies for stroke is the optimal method of cell delivery. NS cells can be transplanted into the brain by intracerebral, intrathecal, intranasal, and intravenous modes of delivery. Of the different possible delivery methods, intra-arterial delivery of NS cells after stroke represents one promising and practical means for NS cell delivery [5]. Intra-arterial delivery offers several advantages compared with other delivery modalities. For example, delivery via an arterial route produces a disseminated delivery of cells in injured regions of the brain, contrasting with the localized delivery of stem cells observed with intracerebral transplantation where a cell bolus is injected into the brain parenchyma.

The feasibility of intra-arterial NS cell delivery is further supported by mechanistic studies showing that NS cells may deploy similar means as immune cells—including chemoattraction, adhesion, and transendothelial migration—to traffic into the brain [6, 7]. Ischemia initiates signaling cascades that make the host brain more permissive for stem cell transplantation. Postischemic activation of the endothelium is accompanied by upregulated expression of various adhesion molecules on endothelial cells including selectins, intercellular adhesion molecule-1 (ICAM-1), and vascular cell adhesion molecule-1 (VCAM-1) [8]. NS cells express the complementary ligands for these adhesion molecules, such as integrins alpha-2, alpha-6, beta-1, and CD49d [6, 9, 10]. Ischemia also triggers astrocytes and microglia to produce chemokines such as CCL2, CCL3, CCL4, CCL5, and CXCL12a [11]. NS cells are competent to respond to such chemoattractant stimuli because NS cells express the complementary receptors CCR1, CCR2, CCR5, CXCR3, and/or CXCR4 [12]. After stroke, endogenous NS cells migrate toward areas of central nervous system injury in CCR2- and CXCR4-mediated manner. These data provide evidence for cross talk between NS cells and immune signaling in stroke [13–15].

The cross talk between the host environment and NS cells also guides transendothelial migration of exogenously delivered NS cells from the vascular into the intraparenchymal space. For example, a critical interaction between VCAM-1 and CD49d promotes trafficking of NS cells into the ischemic brain. Enhancing the CCL2/CCR2 interaction also increases homing efficiency to the ischemic lesion [7]. Understanding the mechanisms by which NS cells home to the brain is likely to lead to improvements in NS cell delivery efficacy [6, 16].

Despite the promise offered by intra-arterial delivery of NS cells, intravenous delivery of cells has also been associated with procedure-related complications, warranting additional studies. In this chapter, I describe an experimental platform for systematically evaluating the intravascular delivery of NS cells into a mouse model of stroke. I provide methods for cell culture of neural stem cells, flow cytometry of neural progenitors, generation of stroke model via unilateral common carotid artery occlusion, intra-arterial injection of stem cells into the brain, behavior analysis, and immunohistochemistry. Identifying experimental parameters critical for the efficacy and safety of intra-arterial delivery will facilitate translation of intravascular NSC delivery from the bench to the bedside.

2 Materials

1. Rat anti-mouse CD49d antibody.
2. Anti-rat secondary antibody.
3. B27 supplement without vitamin A.
4. C17.2 cells.
5. DMEM.
6. DMSO.
7. DNase I.
8. EGF.
9. FGF2.
10. Fetal bovine serum.
11. Glucose.
12. Hanks'-based enzyme-free cell dissociation buffer.
13. Horse serum.
14. L-Glutamine.
15. Neurobasal-A medium.
16. Neutral protease.
17. Papain.
18. PBS.
19. Poly-L-lysine.
20. StemStep Magnetic Bead System.
21. Trypsin/EDTA (0.25%).
22. Neural stem cell medium: Neurobasal-A medium (Invitrogen), 2% B27 supplement without vitamin A, 20 ng/mL fibroblast growth factor (FGF)-2, 20 ng/mL epidermal growth factor (EGF), 1% L-glutamine, 1% antibiotics/antimycotics.
23. C17.2 maintenance medium: DMEM, 10% FBS, 5% horse serum, 1% L-glutamine.

24. Eight percent paraformaldehyde stock: 35 g Paraformaldehyde to 500 mL dH_2O, heat the solution to 65 °C, and stir. Filter and store at 4 °C.

    0.2 M Sodium phosphate buffer, pH 7.4: Monobasic stock, add 25 $NaH_2PO_4*H_2O$ to 1 L H_2O; biphasic stock, add 25 g Na_2HPO_4 to 1 L H_2O. To generate 0.2 M sodium phosphate buffer, combine 800 mL of the monobasic stock with 200 mL of the dibasic stock.
25. Generate 4% paraformaldehyde fixative solution: Combine 8% paraformaldehyde stock with 0.2 M sodium phosphate.
26. pH 7.4 phosphate-buffered saline: 1 L dH_2O, 10 g NaCl, 150 mg KH_2PO_4, 800 mg Na_2HPO_4; then check pH.
27. HEPES buffered Hanks solution pH 7.4: 990 mL dH_2O, 0.15 g $CaCl_2$ 2 parts dH_2O, 2.0 g dextrose/glucose, 2.5 g 10 mM HEPES, 0.25 g KCl, 0.05 g KH_2PO_4, 0.10 g $MgCl_2$ six parts dH_2O, 0.050 g $MgSO_4$ seven parts dH_2O, 7.50 g NaCl, 0.15 g Na_2HPO_4, 90.0 mg NaN_3.

3 Methods

3.1 Donor Cells

3.1.1 Isolation and Cultivation of Neural Stem Cells (Fig. 1)

1. Isolate whole brains from postnatal CCR2+/+ and CCR2−/− mice.
2. Rinse whole brains in PBS.
3. Mince and enzymatically digest whole brains in a cocktail of 2.5 U/mL papain, 1 U/mL neutral protease, and 250 U/mL DNase I in DMEM with 4.5% glucose.
4. Resuspend dissociated cells in neural stem cell medium:
5. Culture as spherical aggregates at 37 °C and 5% CO_2.

Replace half of the culture medium every other day. Omit antibiotics/antimycotics after 10 days in vitro. Passage neurospheres every 5–7 days using Hanks'-based enzyme-free cell dissociation buffer. Use cells from passages 5–10.

3.1.2 Cultivation of C17.2 Neural Stem Cells

1. C17.2 cells were immortalized by infecting neonatal murine cerebellar progenitor cells with avian myc oncogene [17–19]. Coat plates with poly-L-lysine (10 μg/mL) in sterile distilled water.
2. Unlike other cells, feed cells on a weekly basis. Perform half-medium changes with 50% conditioned medium and 50% fresh medium.

3.1.3 Routine Passaging of C17.2 Neural Stem Cells

1. Coat plates or flasks with poly-L-lysine (10 μg/mL) dissolved in sterile distilled water.
2. Add 50 mL of sterile water to 5 mg of poly-lysine.

3. Coat the surface of plates or flasks with 1.0 mL/25 cm^2 (only). Ensure complete coating of the surface.
4. After 5 min, remove the poly-L-lysine solution by aspiration and rinse with sterile water.
5. Allow plates or flasks to dry for at least 2 h before seeding cells and medium.
6. Inspect C17.2 culture before passaging for contamination.
7. Aspirate medium from each well or flask.
8. Add an appropriate volume of trypsin and incubate at 37 °C for 5 min.
9. Remove from 37 °C. For every volume of trypsin, add two volumes of culture medium per well or flask to neutralize trypsin.
10. Transfer cell suspension to a 15 mL conical tube. Centrifuge at 180 × *g* for 3 min to pellet cells.
11. Aspirate the supernatant. Resuspend the cell pellet in medium.
12. If not already removed, aspirate poly-L-lysine from the new plate or flask.
13. Dispense C17.2 cells into the wells or flasks according to the desired passage ratio.
14. Incubate at 37 °C (5% CO_2). Ensure even distribution of cells by moving the plate in several short, back-and-forth, and side-by-side (perpendicular) motions.
15. Visually assess cultures daily to monitor growth until the next passage.

3.1.4 Cryopreservation of C17.2 Cells

1. For cryopreservation, it is desirable to harvest cells growing in the log phase. Approximately 80% confluency is desirable.
2. Aspirate medium.
3. Add 1 mL of trypsin solution and incubate at 37 °C for 5 min.
4. Remove from 37 °C, and inspect visually for detachment.
5. For every volume of trypsin, add two volumes of culture medium per well or flask to neutralize trypsin.
6. Transfer cell suspension to a 15 mL conical tube. Centrifuge at 180 x *g* for 3 min to pellet cells.
7. Aspirate the supernatant. Resuspend pellet in an appropriate volume of DMSO-containing freezing medium. The standard freezing medium is to use 70% culture medium, 20% serum, and 10% DMSO.
8. Pipette 1 mL into each cryovial.
9. Freeze cells and transfer to −80 °C. After 1 day at −80 °C, transfer the cryovial to a liquid nitrogen tank (*see* **Note 1**).

3.1.5 Thawing of Frozen Cells

1. Take frozen vial from liquid nitrogen tank.
2. Thaw in a 37 °C water bath until nearly all vial contents have melted.
3. In a tissue culture hood, wipe newly thawed vial with a tissue soaked in 70% alcohol.
4. Pipette the contents of the cryovial into a tube.
5. Add 5 mL of medium.
6. Determine viable cell density using a hemocytometer.
7. Transfer the appropriate volume of cell suspension to achieve a desired cell seeding density.

Note: Make adjustments to achieve the desired cell seeding density.

3.2 Cell Sorting

Methods described in [20] (Fig. 1) (*see* **Note 2**).

1. Dissociate cells into a single-cell suspension using Accutase.
2. Resuspend cell suspension to a concentration of one million cells per mL in ice-cold PBS supplemented with 10% fetal bovine serum (FBS).
3. Add the appropriate conjugated primary antibody. If dead cell exclusion is desired, propidium iodide can also be added.
4. Incubate for 30 min at room temperature.
5. Wash cells three times and centrifuge at 400 × g for 5 min. Resuspend pelleted cells in 500 μL to 1 mL of ice-cold PBS + 10% FCS. Keep the cell suspension in the dark on ice until FACS.

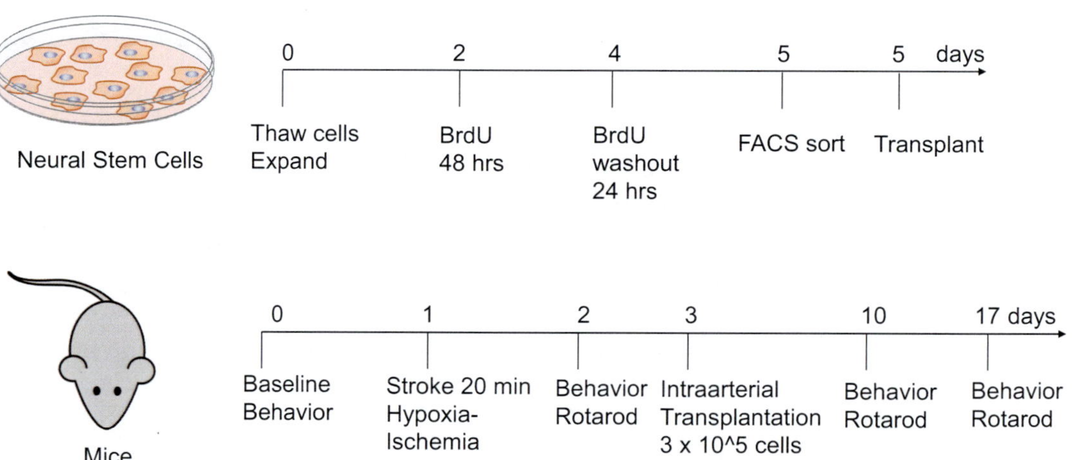

Fig. 1 Study design for preparation of transplanted cells (top) and in vivo transplantation experiments. Adapted from ref. 20 with permission from the journal *Stroke*

6. Conduct positive selection using StemStep magnetic bead system or an alternative magnetic associated cell sorting system (MACS).
7. Stain with PE-conjugated anti-rat secondary antibody or the appropriate secondary antibody.
8. Double-sort using three-laser FACS.
9. After sorting, analyze the purity of CD49d (VLA-4)-positive cells using FACS.

3.3 Stroke Model and Hypoxia-Ischemia

The hypoxia-ischemia stroke model is a two-step process (Figs. 1 and 2). First, block the left common carotid artery; second, transfer to a hypoxic environment. Owed to collateral blood flow from the circle of Willis, occlusion of the ipsilateral carotid artery will not be sufficient to induce an ischemic lesion. However, lowering the levels of oxygen causes the cerebral blood flow to decrease significantly, producing ischemia. By altering the time in which reperfusion occurs, it is possible to "titrate" the level of damage from transient ischemic attack-like phenomenon to larger infarcts. Methods are adapted from [21].

1. Anesthetize mice with an appropriate anesthetic regime.
2. Create a midline neck incision.
3. Isolate the left common carotid artery away from surrounding nerves and apply an aneurysm clip for artery occlusion.
4. Inject 0.5 mL saline subcutaneously to address possible animal dehydration. Apply Lidocaine gel in the wound for pain relief.
5. 2 h of recovery after surgery.

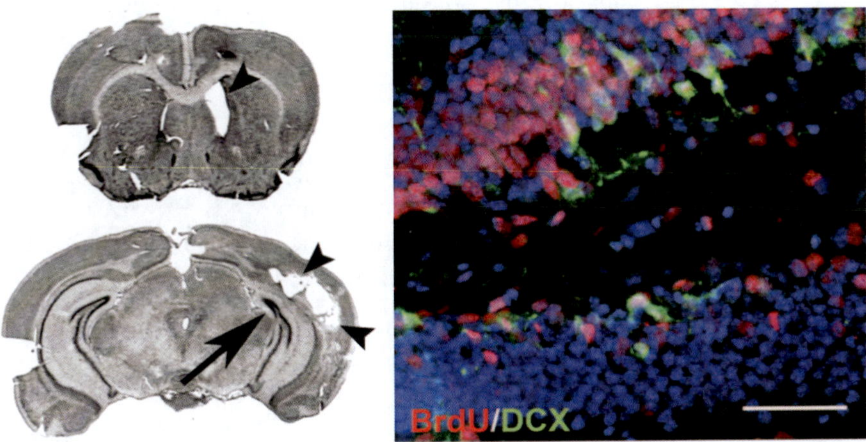

Fig. 2 Stroke model and transplanted NS cells. Left: Two representative Nissl-stained sections. Right: Distribution of NS cells in the mouse brain after stroke delivered through intracarotid injection. Representative confocal image of the hippocampus (BrdU = red; green = doublecortin). Adapted from ref. 20 with permission from the journal *Stroke*

6. Induce hypoxia-ischemia. Place animals in 8% O_2 for 20 min in chambers set at 36 °C.
7. Reperfuse by removing the clip.
8. Close the wound with a clip and store mice in a heated cage for 30–90 min of occlusion.
9. Shorten the remaining sutures and adapt the skin with a surgical suture.
10. Inject 0.5 mL saline subcutaneously again to address possible animal dehydration.
11. Place mice in a heated cage for 2 h to prevent animal hypothermia.

3.4 Neural Stem Cell Transplantation

48 h after stroke, transplant neural stem cells by injecting cell preparations into the left common carotid artery (Figs. 1 and 2) [20, 22].

1. Anesthetize mice with an appropriate anesthetic regime.
2. Create a midline neck incision.
3. Isolate the left common carotid artery away from nerves. To occlude the artery, apply an aneurysm clip.
4. Inject 5 μL of cell preparation (containing 300,000 VLA4-positive cells) into the left common carotid artery using a 10 μL Hamilton syringe. It is advised to use a 33G needle.
5. Close the wound with a clip and store mice in a heated cage for 30–90 min of occlusion.
6. Shorten the remaining sutures and adapt the skin with a surgical suture.
7. To address possible animal dehydration, provide the mice with saline 0.5 mL subcutaneously. For pain relief, apply Lidocaine gel in the wound.
8. Place animals in a heated cage for 2 h to control for body temperature.

3.5 Posttransplantation Behavioral Analyses

The rotarod test is used to evaluate rodent motor coordination (Fig. 1). Acceleration is often employed. However, it is important to emphasize that the rotarod test is not optimized to detect motor skill learning. Methods are adapted from [23] (*see* **Note 3**).

1. Start the rotarod.
2. Place the mouse on the rotarod.
3. Have the mouse to grip the rod.
4. After gripping, start acceleration.
5. Document the speed at which the mouse comes off the rotarod.

6. Record the rotarod speed at the first fall.

To assess behavioral recovery after stroke induction and stem cell transplantation, train mice on the rotarod with accelerating speed starting approximately 3 days before stroke induction. Obtain baseline 1 day before stroke induction. Reassess behavior 2 days after stroke and again before stem cell injection. Generate three groups with poststroke behavior by ranking animals on their rotarod results after stroke. Measure rotarod performance at 10 and 17 days after stroke induction.

3.6 Transcardial Perfusion

Methods for whole-animal perfusion fixation for rodents were adapted from [24] (*see* **Notes 4** and **5**).

1. Administer ketamine/xylazine mixture via i.p. injection to induce anesthesia.
2. Expose the heart by (a) making a small incision below the rib cage, (b) making a small incision in the diaphragm, (c) cutting through rib cage, and (d) lifting the sternum away.
3. Insert a perfusion needle through a cut left ventricle and use a hemostat to clamp the heart and perfusion needle together (*see* **Note 6**).
4. Create an incision on the right atrium.
5. Initiate perfusion first with buffer. The liver should lose color as an indication of good perfusion.
6. Switch to perfusion with fixative. Tremors might be observed.
7. Remove the brain from the head.
8. After isolating the brain, store brain in fixative at 4 °C for 24 h. Then wash the brain three times with PBS. Store the washed brain in PBS.

3.7 Detection of BrdU Pre-labeled Cells Done Using BrdU Staining

BrdU doses ranging between 50 and 200 mg/kg are commonly used for rodents (Fig. 2) (*see* **Note 7**).

3.7.1 BrdU Administration

1. Dissolve BrdU powder in saline to prepare BrdU solution. Filter through a 0.2 μm filter and store.
2. Inject into intraperitoneal cavity.

3.7.2 Standard Immunohistochemical Methods

1. Incubate sections in a blocking solution (it is advisable to use PBS containing 3% fetal bovine serum) for 1 h.
2. Incubate sections overnight at 4 °C in the appropriate primary antibodies diluted in PBS/1% FBS.
3. After several washes in PBS containing 1% serum, incubate sections with secondary antibodies for 1 h. After secondary antibody incubation, wash sections multiple times and stain nuclei with DAPI.

3.7.3 BrdU Staining (Adapted from [25])

1. Place floating sections into 2 M HCl for 2 h at 25 degrees Celsius.
2. Equilibrate sections to pH 7.35. Wash sections four times in PBS. It is advisable to place on a rocker.
3. Block for 1 h (at 4 °C). Again, place on a rocker.
4. Dissolve anti-BrdU antibody in a blocking solution. Incubate sections in the staining solution overnight at 4 °C.
5. Wash sections three times in PBS.
6. Incubate sections with secondary antibody for 2 h at 25 degrees Celsius.
7. Wash sections three times in PBS.
8. Before mounting, remove excess PBS by briefly washing sections in dH_2O.
9. Mount sections on slides.
10. Coverslip.

4 Notes

1. A styrofoam box can be used in a −80 °C freezer for up to 24 h. Then transfer to liquid nitrogen.
2. It is advisable to include serum in buffers to increase viability.
3. Mice can grip the rod rather than walk forward. A rod with a larger diameter may resolve this problem.
4. It is advisable to prepare fixative solution (4% paraformaldehyde) fresh and to use within 72 h.
5. During perfusion surgery, it is important to proceed rapidly as hypoxia-ischemia may cause irreversible changes that may complicate analyses.
6. The perfusion needle should not reach the aorta.
7. BrdU is stable in solution for months when stored at 4 °C.

References

1. CDC Stroke facts. 2017. https://www.cdc.gov/stroke/facts.htm
2. Sinden JD, Hicks C, Stroemer P, Vishnubhatla I, Corteling R (2017) Human neural stem cell therapy for chronic ischemic stroke: charting progress from laboratory to patients. Stem Cells Dev 26:933–947
3. Kelly S, Bliss TM, Shah AK, Sun GH, Ma M, Foo WC, Masel J, Yenari MA, Weissman IL, Uchida N, Palmer T, Steinberg GK (2004) Transplanted human fetal neural stem cells survive, migrate, and differentiate in ischemic rat cerebral cortex. PNAS 101:11839–11844
4. Guzman R, Uchida N, Bliss TM, He D, Christopherson KK, Stellwagen D, Capela A, Greve J, Malenka RC, Moseley ME, Palmer TD, Steinberg GK (2007) Long-term monitoring of transplanted human neural stem cells in developmental and pathological contexts with MRI. PNAS 104:10211–10216
5. Guzman R, Janowski M, Walczak P (2018) Intra-arterial delivery of cell therapies for stroke. Stroke 49:1075–1082

6. Pluchino S, Zanotti L, Rossi B, Brambilla E, Ottoboni L, Salani G, Martinello M, Cattalini A, Bergami A, Furlan R, Comi G, Constantin G, Martino G (2005) Neurosphere-derived multipotent precursors promote neuroprotection by an immunomodulatory mechanism. Nature 436:266–271
7. Andres RH, Choi R, Pendharkar AV, Gaeta X, Wang N, Nathan JK, Chua JY, Lee SW, Palmer TD, Steinberg GK, Guzman R (2011) The CCR2/CCL2 interaction mediates the transendothelial recruitment of intravascularly delivered neural stem cells to the ischemic brain. Stroke 42:2923–2931
8. Frijns CJ, Kappelle LJ (2002) Inflammatory cell adhesion molecules in ischemic cerebrovascular disease. Stroke 33:2115–2122
9. Pluchino S, Quattrini A, Brambilla E, Gritti A, Salani G, Dina G, Galli R, Del Carro U, Amadio S, Bergami A, Furlan R, Comi G, Vescovi AL, Martino G (2003) Injection of adult neurospheres induces recovery in a chronic model of multiple sclerosis. Nature 422:688–694
10. Mueller FJ, Serebryan N, Schraufstatter IU, DiScipio R, Waeman D, Loring JF et al (2006) Adhesive interactions between human neural stem cells and inflamed human vascular endothelium are mediated by integrins. Stem Cells 24:2367–2372
11. Rebenko-Moll NM, Liu L, Cardona A, Ransohoff RM (2006) Chemokines, mononuclear cells and the nervous system: heaven (or hell) is in the details. Curr Opin Immunol 18:683–689
12. Tran PB, Ren D, Veldhouse TJ, Miller RJ (2004) Chemokine receptors are expressed widely by embryonic and adult neural progenitor cells. J Neurosci Res 76:20–34
13. Belmadani A, Tran PB, Ren D, Miller RJ (2006) Chemokines regulate the migration of neural progenitors to sites of neuroinflammation. J Neurosci 26:3182–3191
14. Robin AM, Zhang ZG, Wang L, Zhang RL, Katakowski M, Zhang L et al (2006) Stromal cell-derived factor 1 alpha mediates neural progenitor cell motility after focal cerebral ischemia. J Cereb Blood Flow Metab 26:125–134
15. Yan YP, Sailor KA, Lang BT, Park SW, Vemuganti R, Dempsey RJ (2007) Monocyte chemoattractant protein-1 plays a critical role in neuroblast migration after focal cerebral ischemia. J Cereb Blood Flow Metab 27:1213–1224
16. Friedrich MA, Martins MP, Araujo MD, Klamt C, Vedolin L, Garichochea B et al (2012) Intra-arterial infusion of autologous bone marrow mononuclear cells in patients with moderate to severe middle cerebral artery acute ischemic stroke. Cell Transplant 21(Supp 1):S13–S21
17. Snyder EY, Deitcher DL, Walsh C, Arnold-Aldea S, Hartwieg EA, Cepko CL (1992) Multipotent neural cell lines can engraft and participate in development of mouse cerebellum. Cell 68:33–51
18. Snyder EY, Taylor RM, Wolfe JH (1995) Neural progenitor cell engraftment corrects lysosomal storage throughout the MPS VII mouse brain. Nature 374:367–370
19. Ryder EF, Snyder EY, Cepko CL (1990) Establishment and characterization of multipotent neural cell lines using retrovirus vector-mediated oncogene transfer. J Neurobiol 21:356–375
20. Guzman R, De Los AA, Cheshier S, Choi R, Hoang S, Liauw J, Schaar B, Steinberg GK (2008) Intracarotid injection of fluorescence activated cell-sorted CD49d-positive neural stem cells improves target cell delivery and behavior after stroke in a mouse stroke model. Stroke 39:1300–1306
21. Engel O, Kolodzij S, Dirnagl U, Prinz V (2011) Modeling stroke in mice—middle cerebral artery occlusion with the filament model. J Vis Exp 6:47
22. Rosenblum S, Wang N, Smith TN, Pendharkar AV, Chua JY, Birk H, Guzman R (2012) Timing of intra-arterial neural stem cell transplantation after hypoxia-ischemia influences cell engraftment, survival and differentiation. Stroke 43:1624–1631
23. Deacon RMJ (2013) Measuring motor coordination in mice. J Vis Exp 75:2609
24. Gage GJ, Kipke DR, Shain W (2012) Whole animal perfusion fixation for rodents. J Vis Exp 65:3564
25. Magavi SS, Macklis JD (2008) Identification of newborn cells by BrdU labeling and immunocytochemistry in vivo. Methods Mol Biol 438:335–343

Chapter 11

Ethical Standards for Chimera Research Oversight

Insoo Hyun

Abstract

Ethical standards for stem cell-based chimera research should build on current ethical standards for all animal research, adding stem cell-specific guidance only when necessary to address potential negative effects on animal welfare resulting from heavy biological modification.

Key words Animal research, Animal welfare, Stem cell oversight, Ethics

1 Introduction

For decades, animal chimeras have served as a valuable research tool in diverse areas of biomedical research. Human primary tumor cell lines are commonly transplanted into experimental mice in the course of cancer research. And SCID-hu mouse models of the human immune system have been widely used since the 1980s. Considerations of fairness and justice dictate that one avoids unwarranted stem cell exceptionalism in assessing the permissibility of human-to-animal chimera studies in stem cell research. That is, one should use existing ethical standards for research, unless something specific to stem cell research drives a need for additional ethical standards. Accepting this basic principle of justice (treating like cases alike) means that one should adhere as much as possible to ethical analytic structures used in relevantly similar contexts.

In this chapter I argue that chimera studies involving the transfer of human stem cells or their direct derivatives into animals hosts should be ethically reviewed in a manner that builds on existing animal welfare structures for animal research, and that stem cell-specific standards should be added only to address hypothetical developmental potential and trajectories.

All stem cell-based chimera research involving postnatal live animals is subject to institutional and legal regulations for animal research. Thus, if we are to get a complete picture of chimera

Insoo Hyun and Alejandro De Los Angeles (eds.), *Chimera Research: Methods and Protocols*, Methods in Molecular Biology, vol. 2005, https://doi.org/10.1007/978-1-4939-9524-0_11,

research oversight, we must start with the regulatory requirements for animal research.

Scientific and ethical standards for animal research are well defined in laws, international and national guidance documents, and international standards for voluntary accreditation (for example, the Association for Assessment and Accreditation of Laboratory Care International—AAALAC), as well as in the academic literature [1–3]. In the USA all animal research must be reviewed and approved by an independent scientific and ethics board called an Institutional Animal Care and Use Committee (IACUC). All IACUCs and their equivalent review committees abroad utilize the following standards: (a) the proposed animal research must have substantial scientific merit, and there must be no scientifically acceptable alternative method for answering the research question; (b) the research must be conducted in appropriate facilities by well-trained staff, including veterinarians qualified to care for the species involved; (c) the lowest statistically significant number of animals must be used, without undue pain and stress, and with environmental enrichment appropriate for the species; (d) experimental endpoints must be clearly defined, and euthanasia methods must be humane and veterinarian approved; and (e) ongoing monitoring of approved animal research shall be conducted by the review committee, which has the power to terminate or suspend studies [1]. All of these standards must be upheld before any new animal research proposal can commence. And all top-tier scientific journals require assurance that researchers have met IACUC standards before their work can be published.

In addition to considerations of scientific merit and statistical sample size—which both relate to the scientific integrity requirement of research ethics—the primary focus of animal research oversight is to ensure as much as possible that species-specific animal welfare considerations are responsibly upheld by animal researchers and their institutions. Animal welfare considerations tend to focus primarily on the likelihood of pain, anxiety, and/or suffering experienced by laboratory animals during the course of a study. Importantly, the degree of animal welfare at stake will depend on the physical and mental complexity of the animal, so that a study involving amphibians will have less complicated animal welfare requirements than a mouse study, and a mouse study with less complicated requirements than a study involving pigs, and so forth. Since more complex animals may experience greater degrees of pain and suffering, animal research oversight committees require that the "lowest" possible species be used to answer the research question and that appropriate sedation or anesthesia be used to minimize possible discomfort and distress.

Besides meeting IACUC standards for animal research, all stem cell-based chimera research must also be approved by a Stem Cell Research Oversight (SCRO) Committee. The requirement for

SCRO committee review is stipulated by both American and international stem cell-specific research guidelines [4, 5]. The stated purpose of SCRO review of chimera research is to supplement, not duplicate, IACUC oversight. However, these stem cell research guidelines do not explain in much detail what SCRO committees are supposed to add to the animal research oversight process.

To help fill this void, the Ethics and Public Policy Committee of the International Society for Stem Cell Research (ISSCR) authored an advisory report outlining its stem cell-specific recommendations for the SCRO review of chimera research [6]. The committee presented its recommendations in generally applicable terms, since diverse institutions and international settings may impose their own additional standards for animal research, requiring investigators and reviewers to exercise their best professional judgment in each individual case.

Among the committee's key recommendations (which I merely summarize here) is the proposal that SCRO review should build on existing IACUC standards for animal welfare in scientific research. That is, SCRO committees should add additional ethical standards only if something specific to stem cell research makes it necessary to do so.

Let us consider a primary example of this approach. Current IACUC evaluations of animal welfare are conducted at a species-specific level. This species-specific approach however may still leave open special questions for neurological human/nonhuman stem cell-based chimeras, since such chimerism may (theoretically) pose a change in sentience or behavior from non-chimerized animals. Past experience with genetically altered laboratory animals has shown that reasonable caution might be warranted if genetic changes carry the potential to produce new behaviors and especially new defects and deficits. Best practices today dictate that research involving genetically modified animals must involve the following: (a) the establishment of baseline animal data; (b) ongoing data collection during research concerning any deviation from the norms of species-typical animals; (c) the use of small pilot studies to ascertain any welfare changes in modified animals; and (d) ongoing monitoring and reporting to oversight committees authorized to decide the need for protocol changes and the withdrawal of animal subjects [6].

These four steps aim to help minimize the occurrence of unexpected distress and suffering in genetically modified animals. Contrary to what many laypersons may believe, such bioengineered animals are not usually enhanced above normal species functioning. Rather, the far more typical outcome of genetically modifying research animals is their lower levels of physical functioning (for example, mouse models of cystic fibrosis).

The ISSCR Ethics and Public Policy Committee's report effectively extends these four steps to the SCRO review of human/

nonhuman chimera research. SCRO committees must include experts in stem cell and developmental biology who can assess the likely developmental trajectories produced when signaling modules from different species are mixed during the course of a proposed chimera experiment, taking into account the epigenetic context of regulation in which the mixed cells are going to be deployed. As with genetically modified research animals, SCRO committees ought to require that a baseline of normal animal data be created before chimera experimentation can be permitted. These data might include behavioral and other necessary information for animal welfare, such as hormonal data related to indicators of stress and anxiety. Such baseline data must be grounded in rigorous scientific knowledge or reasonable inferences, not speculative notions about animal mental states and experiences that are contrary to known brain functional requirements and accepted scientific facts.

The committee report also recommends that researchers use small pilot studies to produce initial data on chimeric animals to monitor and describe any deviations from normal behaviors—the same sort of steps researchers would employ for the use of new, genetically modified animals. Contrary to the fears of many opponents of human/nonhuman chimera research, the Ethics and Public Policy Committee was far less concerned about the possibility of "enhanced" humanlike chimeric animals springing into existence in stem cell laboratories. Instead, the committee was much more mindful of the very real possibility (based on experience with transgenic and genetically modified animals) that chimeric research animals will exhibit interrupted equilibria and major biological deficiencies, which would call into action a most diligent application of animal welfare principles.

In summary, the ISSCR chimera report recommends that stem cell investigators, animal research regulators, and SCRO committees extend existing animal welfare standards—in particular, standards for the oversight of heavily modified animals—to cover new human/nonhuman chimeric research animals in a careful, stepwise fashion.

The committee arrived at this series of recommendations after careful consideration of the objections typically mounted against human/nonhuman chimera research. For instance, the committee was not very concerned by the prospect of human chimeric embryo studies. Such research was deemed to be permissible as long as researchers follow ISSCR guidelines by (a) not allowing chimeric preimplantation human host embryos to develop in culture for more than 14 consecutive days, or beyond the appearance of the embryo's primitive streak, and (b) not implanting into a human or nonhuman primate uterus any embryo containing transplanted human pluripotent stem cells. The first stipulation accepts that the moral status of a human or nonhuman host embryo up to

14 days of development is not affected by its possible degree of chimerism. If it is ethically permissible to use entirely *unchimerized* human embryos for in vitro stem cell research, then it should be ethically permissible to use *chimerized* human or nonhuman host embryos for in vitro research. The second stipulation above accepts that it would be ethically undesirable to implant a human/nonhuman chimeric embryo into a uterine environment capable of supporting its full gestation. This restriction is motivated by an interest in avoiding *indiscriminate* human/nonhuman chimerism, especially within a reproductive context that is very hard to predict and control. The restriction is not meant to prevent a more localized and scientifically controlled development of acute human/nonhuman chimeras. After all, both American and international stem cell guidelines allow for the transfer of human pluripotent stem cells into nonhuman animals at their embryonic and fetal stages of development, where the chances of producing acute chimerism are high within localized animal structures. Similarly, the ISSCR Ethics and Public Policy Committee did not categorically advise against the possibility of creating acute human/nonhuman chimeras either in embryonic or later developmental stages. Instead, the report advises that scientifically justified research which may result in significantly humanized mental attributes in human/nonhuman chimeras, while not absolutely prohibited, must be subject to close monitoring and careful data collection pertinent to the welfare of animal subjects.

Finally, the committee was not persuaded by dignity arguments against human/nonhuman chimera research. Some commentators have suggested that chimera research could weaken the human/nonhuman boundary, thereby degrading our human dignity [7, 8]. Similarly, the US National Academy of Sciences states in its stem cell research guidelines that:

> Research in which hES cells are introduced into nonhuman primate blastocysts... should also not be conducted at this time. These kinds of studies could produce creatures in which the lines between human and nonhuman primates are blurred, a development that could threaten to undermine human dignity [4].

Dignity arguments such as these are deeply flawed. The Ethics and Public Policy Committee recognized that human dignity is a multidimensional concept usually characterized by a set of valuable human capacities, such as the capacity for moral agency, self-consciousness, and high-level emotional and cognitive functions. Human dignity is not a property of human cells, but rather a property of whole human beings that is not diminished in the least if new creatures are created that might be said to share some of these capacities. Unlike relational properties, such as being rare or being unique, human dignity is not a kind of property that diminishes or runs out when it is ascribed to more individuals who may deserve a claim to it [6].

I would like to bolster this last point with the following philosophical analysis. Regardless of the specific religious or cultural traditions from which one's notion of human dignity gets its source, human dignity should be understood by all of us as an *intrinsic property* of individuals, not an extrinsic property. An intrinsic property is defined in philosophy as a property that a thing has in virtue of the way that thing itself is. An extrinsic property is that which a thing has which depends on something else outside of itself being true [9]. For example, my being male is an intrinsic property of me, while my being a parent is an extrinsic property. In the case of human dignity, philosophers widely agree that this is something a person possesses intrinsically, regardless of his or her social class, geographical location, cultural milieu, and the like. In fact, it is from this understanding of human dignity (as an intrinsic property of persons) that this notion packs so much political force in the rhetoric of international covenants, such as 1948 Universal Declaration of Human Rights. Human dignity is not something individuals leave at the customs counter when they cross national borders. It is something they carry with them at all times, wherever they are.

To argue that the creation of human/nonhuman chimeras threatens our human dignity is to view human dignity as an extrinsic property, one that can be affected by actions taking place away from us. But it is not an extrinsic property. It is an intrinsic property. Human dignity is violated only when undeserved harms are imposed *directly* onto one's self, such as torture, rape, or slavery. The "undermining of human dignity" argument against human/nonhuman chimera research—be it of an in vitro, embryonic, fetal, or postnatal nature—lacks philosophical coherence when critics claim that experiments taking place outside of ourselves, on a few laboratory animals in scientific research facilities, lessen in any way our own intrinsic properties. Indeed, to make the claim that chimera research violates human dignity is, in my view, an affront to people whose dignity has truly been victimized, such as those ravaged by genocide, or those starving in extreme poverty or famine, or patients in the throes of a degenerative neurological disorder like Huntington's disease. Thus, to avoid being politically glib and insensitive to true human suffering, and for the sake of our own philosophical coherence, we ought to abandon permanently the "undermining of human dignity" argument against chimera research.

In light of the points summarized above, the ISSCR Ethics and Public Policy Committee's recommendations seem to provide a workable framework for overseeing chimera research—building on and remaining consistent with IACUC animal welfare principles, but with added stem cell-specific expertise to consider the further developmental effects on animal welfare of human stem cell-based chimerism.

References

1. National Research Council (1996) Guide for the care and use of laboratory animals. National Academies Press, Washington, DC
2. Orlans FB (1993) In the name of science: issues in responsible animal experimentation. Oxford University Press, New York
3. Rollin BE, Kesel ML (1990) The experimental animal in biomedical research. CRC Press, Boston
4. National Academy of Sciences (2005) Guidelines for human embryonic stem cell research. The National Academies Press, Washington, D.C
5. International Society for Stem Cell Research Guidelines for the conduct of human embryonic stem cell research. 2006. http://www.isscr.org/docs/default-source/all-isscr-guidelines/hesc-guidelines/isscrhescguidelines2006.pdf?sfvrsn=0. Accessed 18 Oct 2018.
6. Hyun I, Taylor P, Testa G et al (2007) Ethical standards for human-to-animal chimera experiments in stem cell research. Cell Stem Cell 1:159–163
7. Kass L (2002) Life, liberty and the defense of dignity: the challenge for bioethics. Encounter Books, San Francisco
8. Schulman A (2008) Bioethics and the question of human dignity. In: Human dignity and bioethics: essays commissioned by the president's council on bioethics. The President's Council on Bioethics, Washington DC
9. Lewis D (1983) Extrinsic properties. Phil Studies 44:197–200

Part III

New Directions and Controversies

Chapter 12

Ethical Considerations in Crossing the Xenobarrier

Sebastian Porsdam Mann, Rosa Sun, and Göran Hermerén

Abstract

Chimeras have been an important part of animal research for decades. Yet crossing the species barrier has always been seen as potentially morally problematic. In recent years, advances in chimeric research and the attendant possibilities—organ xenotransplantation, cognitive enhancement, and others—have given rise to further ethical concern. This contribution surveys the main ethical questions that have been discussed in the literature. We examine two arguments—from the order of nature and from human dignity—which aim to show that chimerization is inherently wrong. Finding the first untenable and the second largely inapplicable, we then turn to two unconvincing arguments designed to show that chimerization must necessarily lead to negative outcomes. Having thus found that no blanket statements can be made on the ethics of chimerization, we examine two important parameters relevant to the ethical evaluation of proposed chimeric research: the argument from moral status and from risk.

Key words Chimeric ethics, Xenobarrier, Moral, Framework, Chimera

1 Introduction

A chimera is "a single organism made up of cells of different embryonic origins" [1]. Chimeras can be formed by a variety of processes, including the mixing of early embryos and the grafting of tissues from different stages of development or from adult organisms. Examples of chimeras in current research include murine models engrafted with human glial cells [2], mouse-rat chimeras with a functioning mouse-derived pancreas [3], pig blastocysts injected with human induced pluripotent stem cells (iPSC) [4], and teratoma formation assays with human embryonic stem cells [5]. Chimeras are different from hybrids and mosaics. Hybrids, such as mules, are the result of breeding between different species. Mosaics are animals carrying genetically different cells originating from the same zygote.

These examples—among many—demonstrate that chimeras can be important model organisms, with applications ranging from establishment of stem cell pluripotency [5] to the study of tumors and the human immune system [6, 7]; organ

Insoo Hyun and Alejandro De Los Angeles (eds.), *Chimera Research: Methods and Protocols*, Methods in Molecular Biology, vol. 2005, https://doi.org/10.1007/978-1-4939-9524-0_12,

xenotransplantation [3, 4, 8]; developmental, neurological/neuropsychiatric and basic science, e.g., [2, 3]. This is by no means an exhaustive list; the potential of chimeric research lies partly in its wide applicability.

2 Preliminary Remarks on Ethics

Ethical problems are related to conflicts of interests or values: How do we decide which, or whose, interests or values should be promoted or favored? Present and future research subjects, animals, researchers, patients, funders, taxpayers, politicians, and other stakeholders do not always have the same short-term or long-term interests. These divergent interests may explain why different answers are given to such questions as follows: Should research be carried out? What type of research? On which conditions, diseases, or systems? Arguments in such debates usually adopt one of the two forms: those that concern knowledge and those that concern values. Is the course of action proposed based on empirical findings or related to values we want to protect or promote? To answer complicated questions of bioethics, both approaches are usually needed.

3 Brief Scientific Background

On the 25th of January, 2017, Hiromitsu Nakauchi's team at the University of Tokyo reported the results of an experiment in which functional mouse-derived pancreata had been successfully grown in mouse-rat chimeras. When these pancreata were removed from the chimeras and transplanted into mice with chemically induced diabetes, not only did the recipients not require immunosuppression after the fifth postoperative day, but also the previously diabetic mice showed normal glucose regulation during a year of subsequent observation [3]. This study proved that, in principle, organ incubation in chimeric hosts followed by xenotransplantation could be successfully carried out without major immunocompromisation.

A day later, Salk Institute researchers published a study involving the injection of human iPSCs into a pig blastocyst. These pig-human chimeric blastocysts were transferred into a surrogate sow and allowed to gestate for 21–28 days. At this point, the researchers confirmed the presence of several established human cell lines in the chimeric blastocyst [4]. This study demonstrated that it may be possible eventually to establish human cell lines in viable pig embryos.

Taken together, these studies suggest that it may be possible to obtain human pancreata, possibly other organs, from pigs—a potentially lifesaving advance given the large gap between human organ demand and donation [8, 9]. Would we be justified in harvesting organs from pigs who cannot consent, like human organ donors must? If so, should we treat the pig donors differently from other pigs? Do we owe these lifesaving pigs anything?

Five years earlier, Steve Goldman's lab engrafted human glial progenitor cells (GPCs) onto the forebrains of neonatal immunodeficient mice. Once matured, the brains of these mice showed widespread integration of human astrocytes and other glial cells. These cells were able to propagate Ca+ signaling threefold faster than the host glia. These human-mice chimeras showed much enhanced long-term potentiation, outperforming their littermates on all four cognitive tests given [2]. This showed that it is possible, in principle, to raise the cognitive powers of animals via chimerization.

By extension, it might be possible to do the same in humans, perhaps through engineering superior neurons for engraftment, perhaps by grafting younger brain tissue onto older brains.

Would either of these possibilities be ethically problematic or permissible? If part-human neural chimeras gain greater cognitive powers, does this affect how we ought to treat them?

In a more recent study, Goldman's lab grafted GPCs obtained from human patients with childhood-onset schizophrenia onto both neonatal myelin-deficient shiverer mice and neonatal myelin wild-type mice brains. Unlike the relatively seamless integration in the earlier study, astrocytes from schizophrenic patient-derived GPCs showed abnormal morphology and delayed differentiation. In addition, GPCs migrated prematurely into the cortex, resulting in less white matter expansion and hypomyelination as compared to controls. The resulting myelin wild-type mice showed several behavioral abnormalities reminiscent of schizophrenic symptoms, including reduced prepulse inhibition, heightened anxiety, sleep disruption, and antisocial traits [10]. This suggests a causal role for glia in schizophrenia.

We might imagine a future of personalized medicine in which tailored chimeric organisms are produced for individual patients, to carry out research on their specific ailments. Is this ethical? What about healthy individuals, wishing to test out the effects of various diets or practices on their own tissue, but housed in a chimeric animal? How severe must the impairment be, or how beneficial the results, to justify the harm caused to these chimeras?

4 Issues and Challenges

There are many challenges and issues, in addition to ethical issues in a more narrow sense. To qualify as an ethical issue it must involve conflicts of interests or values, as already indicated. We may want to identify and distinguish also between scientific, social, and regulatory challenges.

For example, scientific challenges include such questions as follows: Will human organs generated in chimeras be suitable for transplantation? How can we resolve differences in gestation length between humans and other species? Such questions are taken up in other chapters in this volume.

Social concerns—sometimes inspired by religious beliefs, sometimes by secular worldviews, and sometimes by media or culture—may relate to scientific development which do not in any obvious way involve conflicts of interests. We might wonder, for example, whether the mixing of species is inherently wrong, perhaps because it is unnatural, or because it interferes in the divinely ordained order? We address these particular concerns later in the chapter.

5 Ethical and Social Issues

The major ethical and social issues concerning research with and potential clinical use of chimeras include animal welfare (covered in the following chapter and thus not considered here), public discomfort with crossing the xenobarrier, whether crossing the xenobarrier is inherently wrong or violates dignity, risk of unintended consequences, and the effects of humanization on the host organism.

There can be little doubt that chimeric research *is* controversial: the NIH once released a moratorium on federal funding of any research in which human pluripotent cells or neural progenitor cells were placed into any nonhuman vertebrate embryo before gastrulation. Since then, a whole literature examining ethical and social issues has developed. But *why* is it so controversial?

The question that underpins this controversy is whether chimeras are different, in a morally significant way, from other research animals. Most fundamentally, the question is whether the introduction of cells or tissue from other humans or animals is in itself morally wrong, or brings about consequences to the chimera, the environment, or humans in a way that is different from non-chimeric animal research. If this is not the case, then ethical justification for chimeric research will be the same as that for animal research more generally. However, it may be that no blanket statement can be made about the ethical permissibility of chimeric research as such. It is likely that *some*, if not *all*, chimeric research will be morally significantly different from general animal research.

6 Is Crossing the Xenobarrier Inherently or Always Wrong?

We begin by examining whether there might be something inherently wrong with chimerization. Two arguments for this position have been raised in the literature: the argument from the order of nature and the argument from human dignity. In addition, two arguments—a secular version of the argument from the order of nature and the argument from moral confusion—attempt to demonstrate that even if the creation of chimeras is not *inherently* wrong, nevertheless any such attempt is immoral because it necessarily leads to negative consequences. If any of these arguments hold, there will be no need to establish that a subset of chimeric research is morally impermissible. This section presents a critical survey of these positions.

6.1 Argument from the Order of Nature

One of the most frequently invoked, and most frequently dismissed [11], arguments against chimerization is known as the argument from the order of nature. According to this argument, chimerization is wrong because it is unnatural.

The argument exists in at least two forms. The first takes a theological perspective, arguing that chimerization is wrong because it interferes in the divinely ordained natural order. The thought here is that one or more divinities established species and the barriers between them for a reason; humans ought not to take it upon themselves to question or change this establishment. The second, secular version of the argument states that chimerization is wrong because it interferes with the order found in nature. Here it is thought that natural order is the result of a complex evolutionary process which humans could not hope to improve upon and, indeed, are likely to damage if they try.

It should be noted that chimeras exist naturally: fetal cells can sometimes be found circulating in the mother's system; our gut contains billions of microbes, many of which originate in our food or environment. Thus the existence of chimeras cannot be immoral, because unnatural, *tout court*.

A more promising line of argument might be that it is not the *existence* of chimeras which is unnatural, hence immoral, but rather their *creation* by humans. This too fails as a general criterion for the same reason: it is neither unnatural nor wrong to have sexual intercourse leading to pregnancy, and this in turn may lead to chimerization in one sense of the mother. Likewise, this reasoning cannot be rescued by claiming that the creation of chimeras by technological means is unnatural, unless one who holds this position also deems mothers who have become pregnant through artificial means, such as in vitro fertilization, as unnatural and immoral beings.[1]

[1] For those that find this unpersuasive, consider that for decades human-pig chimeras, in the form of humans who have had a pig heart valve implanted, have walked among us.

There are further problems with either conception of the order of nature. If evolution is taken to be the basis for the order of nature, then it cannot be fixed. The processes of survival and adaptation lead to the development of new species and the extinction of others, as well as the modification of currently existing species. On this conception, the argument from the order of nature boils down to the argument that humans should not interfere in this process. But why not? Unless evolution itself possesses moral status or protection, the answer must lie in the consequences of doing so. It might be held that evolution is a fragile process, that too much interference could radically alter the way it works. For example, one might argue that chimeras could outcompete their wild cousins, leading to their extinction and thus to profound changes in the ecosystem. This is a valid argument, but an appeal to consequences cannot support a general prohibition. Humans have interfered with evolution for thousands of years, for example through selective breeding of cattle, dogs, and plants. The outcome has been agriculture and domesticated animals—not an outcome so unequivocally negative as to justify a general prohibition on interference with the nature of plants and animals.

This is also a difficulty if the order of nature is determined by God(s). Presumably, agriculture and husbandry are not against the will of God(s), so why should chimeric animal research be? In order to justify different treatment, we need to specify relevant differences between the actions compared. Furthermore, religiously motivated appeals may be persuasive for those who subscribe to that particular faith, but will carry little weight for those who do not.

The argument from the order of nature requires not only that chimeras be unnatural, but also that chimeras be the kind of unnatural that is inherently wrong. It is clear that not all things unnatural are morally wrong simply in virtue of being unnatural. Artificial lighting, such as electric bulbs, is not found in nature, but it is not morally wrong to produce or consume electricity. John Harris has observed that the entirety of medicine is an attempt to frustrate the course of nature. People naturally get ill and die of various diseases without the intervention of medicine. Nonetheless, it is not morally wrong for someone to be a medical doctor or to benefit from medicine.

However, some things really are morally wrong because they are unnatural. As Robert Streiffer has pointed out, feeding someone unnatural food would be morally wrong because of the negative health consequences that would follow [11]. Similarly, keeping beings in an unnatural environment can be seen as wrong because of the negative consequences that arise from such acts.

Not all chimeras are unnatural. However, some chimeras, such as those with human brain cells, do not exist in nature. But are these chimeras unnatural in the way that medicine is, or in the way that feeding animals unnatural food would be? Chimeras do not automatically lead to negative consequences as unnatural diets or environments do. The consequences of their creation are contingent on what is done to them, or with them, once they do exist. To our knowledge no convincing demonstration has been made that chimeras are inherently wrong because they are unnatural. In the absence of such a demonstration, the argument from the order of nature rests on assertion, and those that do not agree with this assertion are not wrong in dismissing the argument [11].

Importantly, the argument from order of nature fails to distinguish chimeric animal research from non-chimeric animal research. Animal research is itself found nowhere in nature. It may be argued that chimeric animal research is more artificial as it involves the extra step of admixing parts of one animal with parts of another. But non-chimeric animal models are also highly artificial, often having been bred for generations in order to isolate or exaggerate some feature. The argument from the order of nature attempts to specify chimeric animal research as particularly unnatural, but fails to explain why this might be so.

6.2 Moral Confusion

One historically important objection—indeed, the argument that sparked much of the modern debate about chimeras—holds that chimeric research may be "... objectionable because the existence of such beings would introduce inexorable moral confusion in our existing relationships with nonhuman animals and in our future relationships with [chimeras]" [12].

The type of relationship Robert and Baylis point toward is a moral one, and it concerns the way we may treat animals and chimeras. In our culture, animals and humans are treated differently and have different legal status. It is often held, for example, that much scientific research which would be immoral to conduct on humans is nevertheless justified when conducted on animals. Similarly, many believe that it is not wrong to consume animal meat, whereas the consumption of human meat is both a moral offense and a legal crime. Finally, almost everyone would agree that it is better to prefer the death of an animal over that of a human. The justification underlying such differential treatment is that there is a difference in the moral status of humans and animals.

If a being has moral status, it means that they and their interests are morally significant. In other words, to have moral status is to matter morally, to be an entity whose interests must be taken into account in assessing moral questions. Some things—say, a rock—are not moral entities in any normal use of the word; when considering future actions, we do not have to take the interests of a rock into account, because rocks cannot be harmed in a morally meaningful way. By contrast, sentient animals do possess morally relevant interests; they can feel pain, for example, and have an interest in the avoidance of pain and the enjoyment of pleasure as well as a continued existence. Because of this, we cannot treat such animals in the same way we might treat a rock, as their interests give rise to obligations on our part to respect them. It would thus be morally wrong to harm, or to impede those interests [11, 13, 14]. Finally, humans possess not only sentience but also many other features which might give rise to morally important interests, such as the capacities for love, future planning, rationality, and family and friendship ties. These interests give rise to stronger obligations, which in turn is the foundation for the claim that humans have greater moral status than rocks and sentient nonhuman animals [15, 16].

The account given above of moral status is not universally accepted, though as will become clear below in the section devoted to moral status there are strong reasons to prefer it over alternative conceptions. It is one of these alternatives which underlies the argument put forth by Robert and Baylis.

According to Robert and Baylis, there are two types of moral status: contingent and categorical. Contingent moral status describes moral significance that is attributed to a being contingent on the purposes that they serve. For example, the moral status of pets is contingent on being their owner's companions [12]. On this view, the moral status of pets derives not from the obligation to respect their interests, but rather from the obligation to respect the interests of their owners. In other words, Robert and Baylis argue that animals do not have moral status *independently* of the purposes they serve for humans; any moral status that animals may have is *contingent* on the interests humans have in their continued existence. Thus it is not wrong to kill an animal unless it belongs to a human or would in some other way harm the interests of one or more human beings. It follows from this that any and all uses of animals—as food or research subjects, or even as punching bags for those who derive perverse pleasure from torturing animals—are morally neutral (indeed, is not even a moral issue) unless these animals are protected by human interests.

Humans, on the other hand, possess *categorical* moral status, simply by belonging to the category of humans: "human beings have an inviolable right to life simply by virtue of being human" [12]. Only human beings are afforded categorical moral status; their interests and rights are morally protected in society based on this distinction alone. This state of affairs, according to Robert and Baylis, is necessary to justify many socially important institutions and practices, including the slaughter of animals for consumption and their use as research subjects.

Part-human chimeras are problematic for Robert and Baylis's view because they do not quite fit into either category. They do not belong to the category of humans, and therefore do not possess categorical moral status; likewise, they are unlike the other animals, and so might not have merely contingent status. This situation, say Robert and Baylis, leads to inexorable moral confusion. To resolve such confusion, it would be necessary to launch an inquiry into the nature of part-human chimeras, and such an inquiry could only be resolved by establishing the grounds that give rise to moral status. It is likely that such an effort would reveal that the justification given for assuming the contingent moral status of nonhuman animals does not hold water; therefore, argue Robert and Baylis, it might be necessary to prevent chimeric research from happening in the first place "to protect the privileged place of human animals in the hierarchy of being" [12].

Robert and Baylis present this argument as food for thought, not as a position that they themselves endorse. It is not hard to see why: there are several insurmountable problems with this position.

Firstly, Robert Streiffer has pointed out that the most comprehensive survey to date of public opinion concerning the ethics of chimeric research found not a single mention of moral confusion as a reason for discomfort with chimeric research [17]. Since the argument is presented not on its own merits but rather as a concern which others might hold, the absence of empirical evidence is relevant.

Secondly, we do not appear to be confused as to whether chimeras used in current or previous research belong to the category of animals or humans. Neither the SCID-hu mouse—which has a human immune system—nor Goldman's human glia mice raise much doubt as to whether they ought to be considered more human than murine.

The morally important question is whether the principles underlying how we treat animals and chimeras are justified, not whether reassessing their justification is inconvenient. This can perhaps best be appreciated by analogy—what would we think of a similar argument being put forward to justify the continued differential treatment of nonwhite humans, particular religious groups, or women [18]? Questioning the moral status of chimeras is not something to be avoided to protect dubious principles, but rather an "opportunity to reveal ways in which those views are inadequate and make us think about how we might improve them" [19].

6.3 Dignity

The issue of dignity is frequently called upon as a categorical argument against the creation of chimeric beings [20]. Unfortunately, dignity can be a somewhat vague and ambiguous concept admitting of several interpretations, and it is not always made clear which conception is meant by a particular invocation of the concept.

In everyday usage, dignity can refer to distinguish elevated, proper, and graceful ("dignified") conduct from base, licentious, or insipid ("undignified") behavior; alternatively, it can refer to the special status of certain functionaries, such as bishops. In the context of chimeric research, dignity usually refers to one of the two conceptions, outlined below by Karpowicz and colleagues [21, 22]:

1. "Human dignity is a widely shared concept that refers to being worthy or respected because one is human."
2. "Human dignity is based on the recognition that human beings possess, will possess, or have possessed functional and emergent psychological capacities that indicate they are worthy of respect."

The first conception is somewhat similar to the idea of categorical moral status: it is an assertion that humans possess a special quality simply in virtue of their species membership. Indeed, this position must be taken seriously as it is enshrined in international human rights law: the International Bill of Rights, which is a legally binding instrument, recognizes "the inherent dignity and . . . equal and inalienable rights of all members of the human family [as] the foundation of freedom, justice and peace in the world," and the rights contained within these documents "derive from the inherent dignity of the human person" [23–25]. On this view, dignity is possessed by all humans and is the foundation upon which human rights are derived.

As legal instruments, the documents of the International Bill of Rights do not attempt a philosophical justification of this position; rather, they state the consensus view of the hundreds of scholars, theorists, and statesmen (as well as dozens of constitutions, previous rights documents, and much of the then-extant literature) from all over the world who drafted and turned these rights into law [26]. Since this conception is based on enactment in human rights law, it is contingent on developments in politics and international law. This can be seen, for example, in the recent vote of the UK Government to withdraw its Human Rights Act of 1998 (which turned international human rights standards into domestic law). This is also shown in the refusal, to this date, of the United States to recognize the International Covenant on Economic, Social and Cultural Rights. Nevertheless, since human rights law prohibits any experiments which are contrary to human dignity, it is an important question whether chimeric research violates this conception of dignity.

However, efforts since then have been made to ground this conception of dignity philosophically—and thus not contingent on political whim. For example, Morsink has argued that "there exists in the world a realm of objective moral values" which, through the epistemic equipment available to all humans, can be accessed by use of "our shared moral sense, faculty, or conscience" [27]. In essence, this is an argument for a particular version of moral realism, the idea that moral facts are objective and exist independently of humans and theories, and from moral intuitionism: the idea that moral truths can be derived at via intuition and/or are self-evident. It should be noted that the epistemological claim that it is self-evident that dignity belongs to all humans simply in virtue of being human is contentious, and is unlikely to persuade those who do not find this statement self-evident.

Intuitionism aside, the key question here is whether species membership itself can serve as ground for dignity. It has long been pointed out that species membership is not necessarily ethically relevant in and of itself, e.g., [28]. This can be grasped by analogy to other arguments that have attempted to ground special status in group membership alone, such as racism, sexism, or nationalism. Another way to appreciate this is to consider what would happen if we came across nonhuman entities which nevertheless possess many of the features we value and consider uniquely human. Would it make sense to hold that, say, Mr. Spock (of Star Trek fame) should be denied the protections associated with dignity simply because of his Vulcan heritage? In the absence of solid arguments motivating such exceptionalist views, the position that humans possess dignity simply in virtue of being human again rests on bare assertion. Those who do not agree with this assertion are not wrong in dismissing it—and those who do must find reasons to justify this assertion if they wish to have their views considered in rational debate.

Before we move on to the second conception of dignity, it should be noted that there are pragmatic—and by extension, utilitarian—reasons for extending dignity to all humans, though not for denying it to others on the basis of their non-humanity. Throughout history, various “isms” and ideologies have justified torture, murder, and discrimination based solely on group membership, for instance national socialism, social Darwinism, fascism, Marxism-Leninism, Stalinism, and communism. Echoes of this tendency are clearly perceptible even in modern, developed countries in the form of populism, which holds that only some subset of a country or community constitutes its “true” membership and thus deserves recognition and protection [29, 30]. The extension of dignity to all humans can therefore perhaps best be understood as a practical measure to prevent any such future atrocities. This is surely understandable and laudable, but it might be argued that drawing the line at humanity is arbitrary and that other groups, whether chimeric or non-chimeric animals, should also be accorded pragmatic protection lest the deserving be accidentally excluded.

Under the more defensible second explanation, dignity arises from capabilities and psychological capacities similar to those alluded to in the above remarks on moral status. Among the earliest and most influential theorists in this tradition is Immanuel Kant. A full exposition of Kant’s position would be prohibitively lengthy and difficult. But heavily simplified, Kant thought that dignity stems from autonomy, by which he meant the capacity to act in accordance with noninstrumental reason (reason not directed at achieving a particular aim) and the freedom of the will. Autonomous agents in Kant's conception are uniquely able to act on the basis of their own rationality, free from the constraints imposed by rigid determination on, say, the circuits which operate a computer. Unlike a computer, an autonomous agent may choose their own reasons for actions, a capacity which demands respect. Because of this capacity, autonomous agents have *Würde* (moral worth, dignity) and must be respected.

Working in this tradition, Beyleveld and Brownsword proposed a theory of dignity based on agency: defined as being able to carry out, or intend to carry out, a voluntary action for a valued outcome [31]. Applying the moral theory of Alan Gewirth (1978), they argue that agents must respect the basic needs of themselves and other agents, and that it is in this respect that dignity resides. Agents are dignity bearers because they possess capacities and understanding which grant them the ability to define, reflect upon, and pursue interests. Consequently, agents must also value the basic conditions required to fulfill their interests—freedom and well-being in the words of Gewirth. It is crucial that agents assist and not interfere with others' will if everyone is to fulfill these interests; dignity arises from this mutual recognition; rights and duties in turn stem from dignity.

More recently, Griffin has advanced a similar account of dignity based on the capacity to form conceptions of a worthwhile life and the ability to pursue such conceptions in practice. Griffin (2008) refers to this capacity variously as autonomy, agency, or personhood. On his theory, this sort of agency is of ultimate importance for all other valued abilities and therefore constitutes the grounds for dignity.

Common to all these theories is the notion that dignity is the property of being worthy of moral respect because of the capacity to formulate and act in accordance with one's own reason, understanding, and deliberation. With this analysis in mind, we turn to the question of whether dignity, understood in this way, is threatened or violated by chimeric research. Because dignity is a property, it can be stated from the outset that *dignity itself* cannot be violated by chimeric research (an agent either possesses dignity or not; the property itself cannot be affected by experimentation). There are several ways in which dignity (under this conception) might be violated.

One such way is to treat a dignity bearer in ways incompatible with the continued expression of their dignity. Thus, slavery violates the dignity of the enslaved because it does not respect their foundational interest in freedom, necessary for the pursuit of other freely chosen interests. Chimeric research could be a violation of dignity in this sense if and only if the research subject possessed dignity before the experiment (and refused consent) or if one of the outcomes of the research was the conferral of the property of dignity to the subject and, subsequent to this conferral, the subject was not granted the freedom and conditions necessary for the expression of their dignity. Since the vast majority of current chimeras cannot plausibly be said to be dignity bearers, this is not likely to be a live issue.

However, there has been serious debate over whether some great apes, such as chimpanzees, can act autonomously [32]. If this is indeed the case, there would be a strong case for the identification of chimpanzees as dignity bearers; thus, experimentation without consent and not expressly in the interest of the dignity bearer or the interest of others whose well-being is itself an interest of the dignity bearer (e.g., kin) would be a clear violation of their dignity. Similarly, experiments on predominantly human chimeras (e.g., human patients with a pig heart valve) under these conditions would be a violation of their dignity. It further follows that, should it in the future be possible to create dignity bearers through chimerization, it would be a violation of their dignity to continue to treat them as if this conferral had not taken place.

Another way in which dignity can be violated is the removal of the capacities which confer dignity. For example, to cause massive brain injury to a dignity-bearing entity in such a way that the entity is no longer able to act autonomously or in accordance with noninstrumental reason, nor be said to be a person, would be an especially egregious violation, indeed removal, of their dignity. Again, this is unlikely to be an issue for the vast majority of chimeric research; it does, however, raise an important question. It has been pointed out in the context of organ donation that a potential way to source organs without suffering would be to create chimeras, or even non-chimeric humans, genetically engineered never to develop a functioning nervous system [9]. If this were pursued with entities that would otherwise develop the capacities underlying dignity, the question arises whether the *prevention* of the attainment of dignity is itself a violation of dignity. This does not, however, seem to be the case, as similar concerns would be applicable to other entities or materials capable, under the right conditions, of developing into dignity bearers (e.g., sperm, eggs, even skin cells via cloning). Here degrees of potentiality seem of crucial importance: while it is ridiculous to hold that *not* turning a skin cell into a clone is a violation of that (future) clone's potential dignity, it would appear to be a valid concern for entities on the cusp of developing dignity (e.g., through a maturational process). Nevertheless, we can state with confidence that chimeric research involving the destruction of capacities necessary for dignity constitutes a violation of dignity; again, this is only a concern where the chimera had possessed dignity prior to experimentation or attained the property as a result of experimentation.

A third way to violate dignity would be to treat a dignity bearer as if they did not possess the property of dignity, again unlikely in the context of chimeric research. However, this could well be a concern if research is undertaken on entities that might, but do not clearly, possess dignity, including great apes and, more speculatively, elephants, cetaceans, or cephalopods. This would also be a legitimate concern for research which might confer cognitive capacities on chimeras to such an extent that the question arises whether they can be said to possess dignity.

As is evident from these examples, concerns over the violation of dignity can only arise in a very limited subset of all chimeric research. At least in the way dignity is understood here, whole categories of concern—whether the dignity of humanity in general is violated by chimeric research; whether scientists engaged in chimeric research are violating their own dignity; whether the dignity of the human part of a part-human chimera is violated by its admixture with nonhuman elements; and whether dignity itself is violated by chimeric research—are simply nonstarters; they do not make sense under the conception of dignity developed here. Thus, the argument from dignity fails to motivate a general ban on chimeric research, though it is successful in restricting certain kinds of chimeric research. It should also be noted that the argument from dignity does not distinguish chimeric from non-chimeric research: dignity is logically independent of chimerization and must be respected regardless of where it is found.

7 The Argument from Moral Status

Just as dignity is most plausibly conceived as arising from certain capacities, so is moral status—the degree to which an entity's interests matter morally—most plausibly analyzed by reference to the capacities which give rise to interests.

According to the argument from moral status the creation of chimeras with human-type capacities is not wrong in principle, but becomes wrong when such chimeras are not treated in accordance to their status [15]. According to this argument, chimeras with advanced psychological capacities would deserve more moral respect than their non-enhanced counterparts, due to their increased understanding and awareness of their interests and surroundings. Chimeras with sufficiently advanced capacities could, in theory, have interests similar to those of humans. It follows that it would be wrong to continue to treat such chimeras as we currently treat nonhuman animals, for example by using them as research subjects for research that would be unethical to perform on humans.

There is no consensus on which capacities are relevant to moral status. Certainly they include the capacities underlying dignity. In addition, they very plausibly include those underlying personhood: personhood is "associated with a cluster of more specific properties without being precisely analyzable in terms of any specific subset: autonomy, rationality, self-awareness, linguistic competence, sociability, moral agency, and the capacity for intentional action ... A person is someone who has enough of these properties ..." [13]. But other capacities, such as the ability to form kin and friendship bonds, long-term memories, attachments, and plans for the future and to appreciate beauty, gain knowledge, and experience varieties of pleasure and other positive mental states, are also plausible candidates for moral relevance.

It may not be possible to define an exact threshold for moral status, such that those beyond the threshold have full status and those below only partial status. Indeed, according to some researchers moral status may be gradual, such that the more of these capacities an entity possesses, the higher its status [16]. David DeGrazia has argued that we should err on the side of caution in assessing borderline candidates for status [13]. According to his view, some great apes demonstrate enough psychological capacities to call their status into question. Though they are not on the same level as humans, they are also obviously more advanced than some other animals, such as mice. According to DeGrazia, we should treat these apes as if they were persons and thus possess full status (and, on our conception outlined above, also as dignity bearers), so that we are sure that we are not mistreating animals that deserve greater respect. If DeGrazia is correct, similar reasoning would apply to chimeric animals. That is, we should avoid using chimeric animals with capacities sufficiently advanced to call into question the compatibility of their moral status with the kind of research proposed to be carried out on them. Of course this only applies to the kind of research that cannot be ethically carried out without consent on an entity possessing such moral status; there is plenty of research which is ethical to carry out on humans.

The argument from moral status is successful in placing limits, not on the creation of chimeras, but on their treatment and the uses to which they may be put; this is the very meaning of the concept of moral status. Again it should be noted that this is no different from any other research; the moral status of non-chimeric humans and animals must also be respected.

8 Risk

Risk is often defined as a possibility of a negative consequence occurring as a result of an action or event. Risks can be graded in two dimensions: the seriousness of the adverse event and the likelihood or probability of its occurrence. Taking an informed risk requires knowledge or estimates of these consequences prior to performing the action. It is therefore crucial that in creating chimeras, as it would be in the creation of any novel biotechnology with potential to cause great harm and suffering, potential risks are evaluated beforehand so that a more informed decision can be made and risk minimization strategies can be put into place.

In 2012, a paper was published in *Science* [33] announcing the creation of an aerosolized H5N1 avian flu virus. Avian flu viruses have the ability to cause fatal disease in humans, and it was thought prior to this point in time that the absence of their ability to become airborne was the solitary limiting factor toward a global pandemic for some of the more virulent strains. Another group [34] had, in the previous month, reported in *Nature* that the ability to become airborne required the reassortment of viruses (a type of complex genetic exchange). The objective of the paper was merely to establish a ferret model of the research subject virus, as these are frequently used as animal models for human pathogens. Herfst's paper proved that five random mutations caused by transmission among ferrets were sufficient to aerosolize the virus.

Imagine the potential consequences of this experiment: if the virus was accidentally released, mutated to suit a human host, and caused a global pandemic. In 1918, the Spanish flu pandemic was estimated to have killed 50–100 million people [35]; the most recent H1N1 flu epidemic in 2009 caused 100,000–300,000 fatalities [36].

The examination of possible risks associated with the production of this experiment, among dozens of new laboratory models that are created every day, exposes some key elements of risk we ought to contemplate: the intent of performing an action for the greater good, the actual performance of risk analysis in view of foreseeable negative consequences, and risk minimization for unintended consequences.

It is comparatively easier to form an opinion about whether an action is ethically justified when all the contributing factors are well known and recognized. In the face of uncertainty, cost-benefit analyses and utilitarian arguments can be difficult to carry out. The recognition of uncertainty itself, as compared to a known risk, influences decision-making [37–39].

In the context of uncertainty, one tradition, known as the precautionary principle, holds that the proponents of a novel technology hold the burden to prove its safety. Before this burden is met, precaution should prevail. Although this may often be a prudent course to take, the precautionary principle can also be stifling to research, as it can be prohibitively difficult to prove a negative (the absence of risk or uncertainty).

A more useful framework is produced by Hermerén (2012) in a paper addressing the principle of proportionality and its application to bioethical problems. It describes four conditions of any dilemma: importance of objective, relevance of means, most favorable option, and non-excessiveness. These conditions may guide the analysis of decisions involving the creation of chimerized animals, bearing in mind the potential negative consequences (*see* below discussion). A foundation for questions chimeric researchers should be addressing should broadly include the following: Is the objective achievable? Is the use of chimeric animals necessary (proof of superiority over other models should be a bare minimum)? If so how many animals or strains need to be created? Are there any ethical concerns throughout the creation process?

One risk is the potential for off-target chimerism to produce alterations in phenotype. Instead of models of rodents with human immunoglobulins, it may produce autoimmune systems where animal antibodies attack coexisting human antibodies, leading to immunocompromised beings at the mercy of repeated bouts of infections. Something as simple as a single alteration in a crucial protein is all it takes to cause considerable suffering.

Zoonoses are infectious diseases that originate in one species but mutate to infect across species. As in our example of H5N1 influenza virus, spontaneous mutations can give pathogenicity to zoonoses. The spread of zoonotic infections is always initiated by an increase in cross-species contact, for example the ingestion of bush animals such as bats, or mass farming of cattle leading to cutaneous anthrax. Currently, the WHO, FDA, and several national authorities believe the potential benefits of xenograft research to be large enough to justify exploration as long as necessary safety precautions are taken [40].

Early in this chapter, the issue of creation of cognitive capabilities resembling those of humans was raised in the context of moral status and dignity, both fundamental principles in international human rights law and the cornerstone of moral theories. Rights and entitlements aside, let us entertain for a moment what it might be like to experience the world, understand, and remember, but to have little autonomy or liberty to change things. The enhancement of cognitive functions by selectively engrafting neural tissue is not accompanied by physical changes—a true "locked-in" syndrome, an asynchronous mind trapped inside an animal's body understanding the world only by ways of passive perception [22]. For all these reasons, positive steps should be taken to minimize risks for unintended consequences.

9 Conclusions

According to the arguments surveyed here, the ethics of proposed chimeric animal research must be examined on an individual basis. No blanket statements can be made. However, as with any animal research, chimeric research should be carried out only if animal health and welfare are protected according to existing regulations and the risk for adverse effects on the host animal is minimized. In addition, donor cells should be acquired in accordance with relevant current regulations, and research must be examined and approved by an oversight committee, considering the likely patterns of differentiation and distribution of the human cells in the host animal. Unless this is the stated aim of the research, researchers should ensure that the risk of humanization (potentially conferring human cognitive characteristics onto chimeras or creating a human appearance) of the host animal is minimized. Chimeric research can be ethically justified where researchers are transparent concerning uncertainties and current knowledge gaps, when they consider and answer the following four questions, and state the reasons for their answers ...:

1. Is the goal of the research important?
2. Is the contested method necessary to achieve the goal?
3. Can the goal be achieved by using other methods?
4. Are these other methods less controversial or risky?

... and the chimeric animals are treated according to their moral status.

References

1. Nagy, A. and Rossant, J (2001) Chimaeras and mosaics for dissecting complex mutant phenotypes. Int J Dev Biol, 45(3), 577–582. http://www.ncbi.nlm.nih.gov/pubmed/11417901. Accessed 12 Sept 2016
2. Han X et al (2013) Forebrain engraftment by human glial progenitor cells enhances synaptic plasticity and learning in adult mice. Cell Stem Cell 12(3):342–353. https://doi.org/10.1016/j.stem.2012.12.015
3. Yamaguchi T et al (2017) Interspecies organogenesis generates autologous functional islets. Nature 542(7640):191–196. https://doi.org/10.1038/nature21070
4. Wu J et al (2017) Interspecies chimerism with mammalian pluripotent stem cells. Cell 168(3):473–486.e15. https://doi.org/10.1016/J.CELL.2016.12.036
5. Lensch MW et al (2007) Teratoma formation assays with human embryonic stem cells: a rationale for one type of human-animal chimera. Cell Stem Cell 1(3):253–258. https://doi.org/10.1016/j.stem.2007.07.019
6. Rygaard J. and Povlsen CO (1969) Heterotransplantation of a human malignant tumour to "Nude" mice. Acta Pathol Microbiol Scand, 77(4), 758–760. http://www.ncbi.nlm.nih.gov/pubmed/5383844. Accessed 17 Oct 2015
7. Bonyhadl ML, Kaneshima H (1997) The SCID-hu mouse: an in vivo model for HIV-1 infection in humans. Mol Med Today 3(6):246–253
8. Segal BJ, Porsdam Mann S, Mori Y (2018) Partial humanity: the use of chimeric pigs for organ transplantation. HMS Bioethics
9. Shaw D, Dondorp W, de Wert G (2014) Using non-human primates to benefit humans: research and organ transplantation. Med Health Care Philos 17(4):573–578. https://doi.org/10.1007/s11019-014-9565-x
10. Windrem MS et al (2017) Human iPSC glial mouse chimeras reveal glial contributions to schizophrenia. Cell Stem Cell 21(2):195–208.e6. https://doi.org/10.1016/J.STEM.2017.06.012
11. Streiffer R. Human/non-human chimeras. Stanford Encyclopedia of Philosophy Fall. 2014. http://plato.stanford.edu/entries/chimeras/. Accessed 6 Nov 2015
12. Robert JS, Baylis F (2003) Crossing species boundaries. Am J Bioeth 3(3):1–13. https://doi.org/10.1162/15265160360706417
13. DeGrazia D (2007) Human-animal chimeras: human dignity, moral status, and species prejudice. Metaphilosophy 38(2–3):309–329.

https://doi.org/10.1111/j.1467-9973.2007.00476.x
14. Buchanan A (2009) Moral status and human enhancement. Philos Public Aff 37 (4):346–381. https://doi.org/10.1111/j.1088-4963.2009.01166.x
15. Streiffer R (2005) At the edge of humanity: human stem cells, chimeras, and moral status. Kennedy Inst Ethics J 15(4):347–370
16. DeGrazia D (2008) Moral status as a matter of degree? South J Philos 46(2):181–198. https://doi.org/10.1111/j.2041-6962.2008.tb00075.x
17. Streiffer R (2003) In defense of the moral relevance of species boundaries. Am J Bioeth 3 (3):37–38. https://doi.org/10.1162/15265160360706543
18. Savulescu J (2003) Human-animal transgenesis and chimeras might be an expression of our humanity. Am J Bioeth 3(3):22–25. https://doi.org/10.1162/15265160360706462
19. Bok H (2003) What's wrong with confusion? Am J Bioeth 3(3):25–26. https://doi.org/10.1162/15265160360706471
20. Palacios-González C (2015) Human dignity and the creation of human-nonhuman chimeras. Med Health Care Philos 18 (4):487–499. https://doi.org/10.1007/s11019-015-9644-7
21. Karpowicz P, Cohen CB, van der Kooy D (2004) It is ethical to transplant human stem cells into nonhuman embryos. Nat Med 10 (4):331–335. https://doi.org/10.1038/nm0404-331
22. Karpowicz, P., Cohen, C. B. and van der Kooy, D (2005) Developing human-nonhuman chimeras in human stem cell research: ethical issues and boundaries. Kennedy Inst Ethics J, 15(2), 107–134. http://www.ncbi.nlm.nih.gov/pubmed/16149204. Accessed 20 Oct 2015
23. United Nations The universal declaration of human rights. 1948. http://www.un.org/en/documents/udhr/. Accessed: 25 May 2015
24. United Nations International covenant on civil and political rights. 1966a
25. United Nations International covenant on economic, social and cultural rights. 1966b. http://www.ohchr.org/EN/ProfessionalInterest/Pages/CESCR.aspx. Accessed 25 May 2015
26. Morsink J (1999) The universal declaration of human rights: origins, drafting, and intent. University of Pennsylvania Press, Philadelphia
27. Morsink J (2009) Inherent human rights: philosophical roots of the universal declaration, Pennsylvania studies in human rights. University of Pennsylvania Press, Pennsylvania
28. Singer P (1995) Animal liberation. HarperCollins, Pimlico
29. Glover J (2000) Humanity: a moral history of the twentieth century. Yale University Press, New Haven
30. Müller J-W (2016) What is populism? University of Pennsylvania Press, Philadelphia
31. Beyleveld D, Brownsword R (2001) Human dignity in bioethics and biolaw. Oxford University Press, Oxford
32. Beauchamp TL, Wobber V (2014) Autonomy in chimpanzees. Theor Med Bioeth 35 (2):117–132. https://doi.org/10.1007/s11017-014-9287-3
33. Herfst S et al (2012) Airborne transmission of influenza A/H5N1 virus between ferrets. Science 336(6088):1534–1541. https://doi.org/10.1126/science.1213362
34. Imai M et al (2012) Experimental adaptation of an influenza H5 HA confers respiratory droplet transmission to a reassortant H5 HA/H1N1 virus in ferrets. Nature 486(7403):420–428. https://doi.org/10.1038/nature10831
35. Taubenberger JK et al (2012) Reconstruction of the 1918 influenza virus: unexpected rewards from the past. MBio 3(5): e00201–e00212. https://doi.org/10.1128/mBio.00201-12
36. Dawood FS et al (2012) Estimated global mortality associated with the first 12 months of 2009 pandemic influenza A H1N1 virus circulation: a modelling study. Lancet Infect Dis 12 (9):687–695. https://doi.org/10.1016/S1473-3099(12)70121-4
37. Sahlin N, Persson J, Vareman N (2011) Unruhe und Ungewissheit: stem cells and risks. In: Hermerén G, Hug K (eds) Translational stem cell research: issues beyond the debate on the moral status of the human embryo. Human Press, Springer, Heidelberg, pp 421–429
38. Sahlin N (2012) Unreliable probabilities, paradoxes, and epistemic risk. In: Roeser S et al (eds) Handbook of risk theory. Springer, Heidelberg, pp 477–498
39. Kortenkamp KV, Basten B (2015) Environmental science in the media. Sci Commun 37 (3):287–313. https://doi.org/10.1177/1075547015574016
40. Fishman JA, Scobie L, Takeuchi Y (2012) Xenotransplantation-associated infectious risk: a WHO consultation. Xenotransplantation 19 (2):72–81. https://doi.org/10.1111/j.1399-3089.2012.00693.x

Chapter 13

Neurological Chimeras and the Moral Staircase

Daniel Counihan

Abstract

The possibility of human/nonhuman acute neurological chimeras stirs much controversy. There is concern that neurologically humanizing a research animal could grant it protected moral status, making the performance of invasive experiments on the resulting chimera unethical. However, the real ethical problems may involve smaller increases in moral status along what may be termed the "moral staircase" of research subjects, both animal and human. Through appropriate recognition of incremental increases in moral status along the moral staircase, acute neurological chimera experiments may be permitted to ethically proceed. The real ethical issues around human/nonhuman acute neurological chimeras lie in the realm of animal welfare.

Key words Neurological chimeras, Moral status, Animal welfare, Humanization

1 Introduction

Chimera research has the potential to cause much public concern. In the United States there exists a moratorium on federal funding for certain types of stem cell-based interspecies chimeras, making research difficult [1]. Yet, through private research and nonfederal funds, much of this research has continued to develop. What is it about chimeras that makes them so controversial?

It seems that major concerns revolve around the possibility of generating human/nonhuman acute neurological chimeras—that is, of biologically humanizing large portions of the central nervous systems and/or brains of animal hosts. Four main areas of concern are (a) moral taboos, (b) species mixing, (c) human dignity, and (d) biological humanization that leads to the moral humanization of neurological chimeras. However, none of the concerns appears to be founded in a reality.

I argue in this chapter that the real ethical concerns surrounding human/nonhuman acute neurological chimeras lie in their moral status as research tools and in animal welfare. Potential alterations in the cognitive capacities of neurological chimeras

Insoo Hyun and Alejandro De Los Angeles (eds.), *Chimera Research: Methods and Protocols*, Methods in Molecular Biology, vol. 2005, https://doi.org/10.1007/978-1-4939-9524-0_13, © Springer Science+Business Media, LLC, part of Springer Nature 2019

might increase their moral status as objects of research but only in small, limited amounts. Regulation and oversight are necessary to make sure that any changes in moral status are recognized and adequate steps are taken to address possible increases in such status.

2 Approaches to Moral Status

Human beings have rights and deserve respect. And animals should not be harmed without moral justification. These two broad claims are grounded in an overarching belief that humans and animals are beings with moral status. Philosophers commonly view a being as having moral status "if and only if it or its interests morally matter to some degree for the entity's own sake" [2]. A being with moral status can be wronged, unlike other entities (cars, for example) that can be treated badly without being wronged. Moral status also has differing degrees of actualization [3]. This means that higher degrees of moral status offer more protections and rights than lower degrees. At the highest level is full moral status which is typically granted to humans (and possibly to chimpanzees as well) [4]. However, to get to any determination of moral status at any level, the *grounds* for ascribing moral status must first be considered.

One common view for the grounds of moral status equates it with an entity's actually (not just potentially) having some particular kind of sophisticated cognitive ability [2]. Although while this specific ability is not always agreed upon by those who adhere to this general approach, such a cognitive view maintains that moral status is grounded in an entity's *actually* being able to do something important cognitively. However, this approach to moral status comes with a few issues. For example, if we use this approach to ascribe to adult humans full moral status because of their complex (yet-to-be-philosophically-defined) cognitive abilities, then a developing child or baby, or even a mentally disabled adult, may not meet the requirement for having full moral status. Yet, from a commonsense view, we still offer full moral status protections to all humans after birth [5]. This example highlights that requiring an actual complex level of cognition would not give all humans a clear ground for having full moral status.

Another common view defines the grounds of moral status as one's having the *potential* to have a sophisticated cognitive ability [6]. This view would bring normally functioning children and babies into the class of beings that possess full moral status, since the former have the potential for sophisticated cognitive function. This approach however also carries problematic implications. For example, it can be overly inclusive, bringing first-trimester fetuses in utero and even preimplantation human embryos into the realm of full moral status, which many believe to be an argument resting on a philosophically problematic nonsequitur [3]. Just because entity X could potentially become entity Y (under the right circumstances, many of which are social in nature and thus not guaranteed by any means) it does not logically follow that X *now* has the exact same status as Y. This view is also underinclusive, stripping rights and protections away from the severe cognitively impaired, which the commonsense view includes.

Furthermore, with regard to human/nonhuman acute neurological chimeras, it is uncertain at best whether any chimeras would have the potential to acquire new mental capabilities—let alone those that approximate the cognitive ability humans need for higher moral status. So even if one were to embrace a "potentiality" approach to moral status, knowing whether this approach applies to neurological chimeras is a separate problem. The first step would be to determine that a specific chimera has developed a new cognitive ability; second, to determine if this means it has an increase in moral status; and third, to treat that particular chimera as having the potential for that cognitive ability. This view is limited as it must wait until a cognitive ability arises to raise moral status. Furthermore, once it is determined that a chimera of a particular variety of chimeras in an experiment has a new cognitive ability, other chimeras of that type must be individually assessed before any general conclusions could be drawn about whether all chimeras of that sort can have similar increases in moral status.

3 Degrees of Moral Status and Their Adjustments

As mentioned earlier, moral status admits of degrees, with human beings typically viewed as occupying the top spot. There are three leading types of theories on the moral status of beings that do not necessarily qualify for full moral status. The first are indirect theories, which offer nonhuman beings no protections as a matter of morality [7]. The common idea here is that nonhuman beings cannot be morally wronged. This type of theory is championed by Kant, Aquinas, and a few others [7]. However, these views fail to consider that we still offer protections to animals for the animal's benefit, not ours. Limitations include protections against animal abuse and against causing an unnecessary pain during research. These protections form much of the basis for the Three Rs in animal research: reduce, refine, and replace [8]. These protections exist for the sake of animals themselves.

The second type of theory is moral equality theories, which suppose that animals have the exact same moral status as humans do, and thus should be offered the same protections [7]. These theories are championed by Singer, Regan, and animal rights groups [7]. The issue with this type of theory is that, even though we may afford animals some rights, we normally do not believe that they possess moral status that is equal to humans. Even the most extreme animal rights groups would probably agree that it is better to kill a mouse than a human, since a human loses more in death [9].

The third type of theory is unequal indirect rights theory. On this view animals can have at least some level of moral status, and different animals can occupy different levels of moral status [7]. One might imagine the varying possible degrees of moral status as a moral "staircase." At the top of the staircase is full moral status where humans sit. At the bottom are entities like single-celled organisms, where no moral status is granted. In between is a series of steps. Each step signifies greater protections and more moral weight. There are two options for this moral staircase. The first option is that there is just one step between humans and single-celled organisms where all other animals reside. On this option, all nonhuman animals would all be offered the same protections and moral weight. A second option is a series of steps. At the lower steps are lower animals, such as fish and mice. At the upper steps are animals of great complexity: dolphins, chimpanzees, and other higher order primates. The higher up the moral staircase one goes, the more protections one is offered. In the context of research ethics, these different protections can mean different housing requirements, limits on the number of subjects, or stronger reasons not to use more complex animals at a particular point along the moral staircase but to use a species lower down on the hierarchy. This seems to be the idea behind, in large part, the Three Rs of animal research ethics. Replacement calls for using the most basic animal sufficient to answer one's research question instead of a higher animal, like choosing a rodent model over nonhuman primates, or replacing the use of animals entirely with nonliving research models where possible.

In light of current animal research guidelines and practices, it seems that our current system of animal research is based on a threshold view where the moral status differences between research animals come down to differences in their mental capacities or sentience [9–11]. Implicit in this is a threshold view of moral status, which allows for simplicity in the level of protections we offer to classes of animals. In addition, the grouping of animals onto different steps on the moral staircase allows for simplicity between closely related species. There is a risk of not respecting the actual differences between species; however the level of protections offered by the regulations around animal testing seems to err on the side of more protections instead of less.

In applying moral status frameworks to human/nonhuman acute neurological chimeras, there may be special risks to consider. But we must be careful not to overestimate these risks. As discussed earlier, the potential to have complex cognitive capacities cannot be readily assumed when dealing with chimeras. While we might be able to increase the mental capacities of neurological chimeras, the risk of increasing them to the level of humans, granting them full moral status, seems very unlikely. A more likely risk, if there is such a risk at all, is increasing their cognitive abilities just slightly, but not to the level of humans. This has the potential to increase their moral status up the moral staircase without increasing their status to the fullest level of humans. Take, for example, a neurologically humanized mouse elevated up the moral staircase to the level of a cat. The moral risk here would be to raise this chimeric mouse to the moral status of a cat and then to continue to treat it like a normal mouse [12]. Increasing the moral status (i.e., cognitive complexity and mental capacities) of chimeric animals without proper and commensurate recognition and change in how the chimeric animals are treated in the course of research could increase the suffering and therefore the potential for morally wrong research via inadequate animal welfare protections. To go back to the chimeric mouse with the newfound status of a cat example, some experiments one could do on a mouse would be unethical to do on a cat, including lack of attention to new environmental enrichment requirements. In short, the most likely ethical risk of creating human/nonhuman acute neurological chimeras is the risk of failing to apply appropri ate standards of animal welfare warranted by their new capabilities.

4 Movement Along the Moral Staircase

When dealing with the moral staircase implicit in all animal research, the onus for changes in moral status lies with the researcher. This will require researchers to be aware of general behaviors of the chimeric host animals as well as general behaviors of animals at higher steps. If researchers notice an abnormal behavior, it is their responsibility to stop performing experiments so that a review committee can determine if the chimeric animal is exhibiting a behavior that would indicate an increase of moral status. If their moral status is increased, then the experiments are not necessarily unethical as long as researchers give animal welfare protections worthy of the increase in moral status. The rest of this section provides a very general model for the steps, the typical behaviors at the steps, and a few behaviors to be aware of that might indicate an increase in moral status. This section demonstrates a way of thinking about potential increases in moral status. A list of general protections for different animals in research can be found in the IACUC Policy on Animal Housing and Enrichment [13]. This general list was used to inform the following differences in moral status levels and the key protections added at each step.

4.1 Step 0

The bottom of the staircase consists of entities with minimal to no moral status. This would include entities like single-celled organisms and other non-sentient living things. Lack of sentience is the defining characteristic of this group at Step 0. There are few if any regulations around these entities. Major regulations that do exist are not based on the impact the research will have on these entities, but rather on the effect research could have on other beings with moral status. For example, when working with infectious disease pathogens, regulations exist not for the sake of the pathogens themselves but for the effects that the pathogens could have on the outside world.

4.2 Step 1

Step 1 includes small vertebrates like mice, rats, and other like species. For this classification of research animals, the defining characteristic is sentience. Sentience as the moral basis for this step is defined as "feeling or sensation as distinguished from perception and thought" [14]. This can include the ability to feel pleasure and pain [10, 11]. This ability to feel pleasure and pain has given rise to certain animal welfare protections and regulations around beings with this capability.

It is unlikely that sentience is a property that could be gained solely from chimera research alone, since all host species for chimera research are already sentient. In other words, there should be no emergence of sentience where sentience did not already exist during the course of human/nonhuman acute neurological chimera research.

4.3 Step 2

The next step on the moral staircase includes companion animals, such as cats and dogs. The defining characteristic for these animals appears to be the need for positive interaction from humans [15]. These animals need positive interaction for quality-of-life purposes. Surveys and assessments are used on a daily basis in veterinary practice to assess animals and to inform researchers of best practices [16]. As researchers regularly interact with their research subjects, determinations of quality of life may be hard to measure. Performing regular quality-of-life assessments with animals at this step and looking for abnormal behaviors is the key. If positive interactions raise both the quality-of-life assessments and diminish abnormal behaviors, researchers need to consider whether chimeras arising from host animals at Step 2 have had a change in moral status.

4.4 Step 3

Large mammals, such as agricultural animals, make up the next step in the moral staircase. This level involves a constant need for socialization, shown in regulations by the need for a companion of their own species. This can be accomplished by having another like-species animal in the pen to cohabitate with or a mirror to replace the companion [13]. In research this has been shown to be beneficial by looking at animals' reactions to stress responses when left alone. In pigs, research has shown increased cortisol levels well above the normal stress response [17]. In sheep, looking at just the stress response without many biological measures showed a similar heightened stress level when kept in isolation [18]. This constant need for socialization is seen in other large species.

One key characteristic to look for in a chimeric animal is depression or abnormal behavior when the animal is isolated. Abnormal behaviors may sometimes be seen after a surgery when an animal needs to be isolated [15]. Importantly, abnormal behavior can be due to the surgery itself. Whenever an abnormal behavior is seen in an isolated chimeric animal, researchers should take healthy animals that did not go through the surgery, isolate them, and assess whether a similar abnormal behavior develops. If not, the chimera's abnormal behavior is likely due to surgery. If the behavior does develop, then it is up to the researchers and a review committee to investigate further any mental changes that might affect the moral status of the chimeric animal.

4.5 Step 4

At level 4 on the moral staircase are nonhuman primates. Nonhuman primates are used sparingly in research due to the need to follow the Three Rs. As research animals, nonhuman primates have all of the protections of the previous three steps including sentience, positive interactions, and constant socialization. They do not quite reach the step above this—almost, but not quite—to the point of human protections. Currently they are generally used in immunology and neurological functioning research [19].

It is their increased cognitive functions that make their use in invasive research controversial. In their positive interactions with chimeric nonhuman primate research animals, researchers can regularly assess their cognitive capabilities, remaining aware of normal behaviors in unaltered nonhuman primates and looking for major changes.

4.6 Step 5

The fifth step on the moral staircase is for human beings. The defining characteristics for this step are self-consciousness and personhood. These characteristics seem to require social learning and are not automatically guaranteed by mere biological properties [20].

The protections gained at this level are substantial. Informed consent and communications of the research risks and benefits are just a few of the protections specific to this final step along the moral staircase. There are many more regulations, protections, boards, certifications, etc. that must be gained to do research with humans. We as human beings stand as an example of what full moral status confers, and research with beings at this level should be taken with the highest precautions.

5 Summary and Recommendations

The steps in the moral staircase sketched above are meant to be offered as a practical starting point in thinking about the ethics of chimera research. It is a highly theoretical model that needs more research and consideration by research boards to come up with a list of proper behaviors and protocols. That being said there are a few major takeaways that can help inform future research on chimeric animals.

First, the onus is on researchers to consider and measure changes in chimeric animals that would place them on a different step along the moral staircase compared to unaltered hosts. If such changes occur, researchers must respond accordingly. It would be unethical to treat a chimeric animal like an unaltered host if in reality it has gained a cognitive ability or trait that would place it on a higher level of research animal protections.

Second, current oversight of animal research is already extensive. In the United States, Institutional Animal Care and Use Committees (IACUCs) aim to ensure animal welfare as well as safety of researchers, institutions, and the public [15]. Chimera experiments using stem cells must go through an additional committee called the Stem Cell Oversight Committee (SCRO) [21]. This committee's role is more specialized than that of the IACUC, since the former is charged with overseeing stem cells and their use in research [21]. But these two committees, as they are currently constituted, may not be enough. Somehow there needs to be an additional assessment on whether neurological chimeras gain cognitive capacities that could change their status along the moral staircase.

A key recommendation that follows from the discussions above is the need for an animal behavior expert who can serve as a research consultant for the institution. Such an expert would allow for constant monitoring of chimeric animals in order to make recommendations to chimera research oversight committees about what to do with human/nonhuman acute neurological chimeras when abnormal behaviors are observed.

The conversation surrounding human/nonhuman acute neurological chimeras should focus on the animal welfare impacts of altering host species. The moral status of research animals is grounded in their sentience and other basic mental capabilities. Changes in the mental capabilities of chimeric animals, and by extension their moral status, will have to be appropriately recognized. Without recognition of these factors, neurological chimera research might inadvertently violate animal welfare standards in research.

Through appropriate oversight, the risks outlined above can be minimized. If abnormal behaviors are discovered, researchers and oversight committees should try to assess any changes in moral status. Once these changes are determined, researchers should develop protocols that align the treatment of chimeric animals to their nearest relevant step along the moral staircase.

References

1. Pullen LC (2016) NIH to lift moratorium on chimera research funding. Am J Transplant 16:3311–3312. https://doi.org/10.1111/ajt.14094
2. Jaworska, A, Tannenbaum, J. The grounds of moral status. In: Stanford encyclopedia of philosophy. 2018. https://plato.stanford.edu/entries/grounds-moral-status/. Accessed 6 Feb 2018.
3. Green RM (2002) Stem cell research: a target article collection: Part III—determining moral status. Am J Bioeth 2:20–30
4. Coates A. Is this chimpanzee a non-human person? In: The independent. 2017. https://www.independent.co.uk/news/long_reads/is-this-chimpanzee-a-non-human-person-a7941876.html. Accessed 27 Mar 2018.
5. United Nations (UN) (1989) Convention on the rights of the child. 1577 UNTS 3
6. Stone J (1987) Why potentiality matters. Can J Philos 17:815–829
7. Wilson SD. Animals and ethics. In: Internet encyclopedia of philosophy. 2018. https://www.iep.utm.edu/anim-eth/#H2. Accessed 15 Apr 2018
8. Fenwick N, Griffin G, Gauthier C (2009) The welfare of animals used in science: how the "Three Rs" ethic guides improvements. Can Vet J 50:523–530
9. DeGrazia D (2008) Moral status as a matter of degree? South J Philos 46:181–198
10. Douglas T (2013) Human enhancement and supra-personal moral status. Philos Stud 162:473–497
11. Biller-Andorno N (2002) Can they reason? Can they talk? Can we do without moral price tags in animal ethics? In: Applied ethics in animal research: philosophy, regulation, and laboratory applications. Purdue University Press, West Lafayette, IN
12. Streiffer R (2005) At the edge of humanity: human stem cells, chimeras, and moral status. Kennedy Inst Ethics J 15:347–370
13. IACUC Policy on Animal Housing and Enrichment University of California: Irvine. 2015. https://research.uci.edu/compliance/animalcare-use/research-policies-and-guidance/environmental-enrichment.html. Accessed 25 Aug 2018
14. Merriam-Webster. Sentience. Merriam Webster incorporated. 2018. https://www.merriam-webster.com/dictionary/sentience. Accessed 2 Sept 2018
15. National Research Council (2011) Guide for the care and use of laboratory animals, 8th edn. National Academy Press, Washington, D.C
16. Yeates J, Main D (2009) Assessment of companion animal quality of life in veterinary practice and research. J Small Anim Pract 50:274–281
17. Barnett JL, Cronin GM, Winfield CG (1981) The effects of individual and group penning of pigs on total and free plasma corticosteroids and the maximum corticosteroid binding capacity. Gen Comp Endocrinol 44:219–225
18. Price EO, Thos J (1980) Behavioral responses to short-term social isolation in sheep and goats. Appl Anim Ethol 6:331–339
19. Sternudd K. Research with monkeys. 2018. https://ki.se/en/research/research-with-monkeys. Accessed 1 Sept 2018
20. Hyun I (2013) The limit case—acute human-nonhuman neurological chimeras. In: Bioethics and the future of stem cell research. Cambridge Univ. Press, New York
21. Hyun I (2013) Pegasus and the chimaera—stem cell-based animal research. In: Bioethics and the future of stem cell research. Cambridge Univ. Press, New York

Chapter 14

Isolation, Cryopreservation, and Transplantation of Spermatogonial Stem Cells

Nilam Sinha, Eoin C. Whelan, and Ralph L. Brinster

Abstract

Spermatogonial stem cell (SSC) culture and transplantation pave the way for clinical restoration of fertility in male prepubertal cancer survivors. In this chapter we detail the steps for isolating and freezing testicular tissue along with protocols for the subsequent recovery from cryopreservation and transplantation of cells into a recipient testis. Transplantation of cultured or thawed SSCs provides not only a functional assay for identification of stem cells, a critical tool for the study of the germline stem cell niche in model organisms, but also a framework for reconstitution of spermatogenesis in humans. As proof of concept, the outlined methods have been performed successfully in the murine model and have the potential to be translated to clinical environments.

Key words Spermatogonial stem cells, Germ cell, Testicular biopsy, Cryopreservation, Transplantation, Microinjection, Spermatogenic colony, Infertility, Prepubertal cancer

1 Introduction

Spermatogonial stem cells (SSCs) are the foundation of male reproduction; the balance of self-renewal and differentiation is vital for male fertility. As a result, SSCs have been an area of intense research, particularly regarding modification of the male germline and regeneration of spermatogenesis. The potential for prospectively preserving the germline is of critical importance when considering infertility caused by cancer treatment of prepubertal boys. Advances in therapeutics have led to high cancer survival rates in children. Pediatric patient survival rate has progressed from a 58% 5-year survival rate in 1975–1977 to an 83% survival rate in 2001–2007 [1]. However, the side effects of these treatments often result in reduced reproductive success or total infertility [2]. Because prepubertal boys do not produce sperm that could be banked, a treatment involving testicular biopsy, culture, and cryopreservation of their SSCs would allow for future reintroduction of the cultured cells into the testis and restoration of fertility.

Insoo Hyun and Alejandro De Los Angeles (eds.), *Chimera Research: Methods and Protocols*, Methods in Molecular Biology, vol. 2005, https://doi.org/10.1007/978-1-4939-9524-0_14, © Springer Science+Business Media, LLC, part of Springer Nature 2019

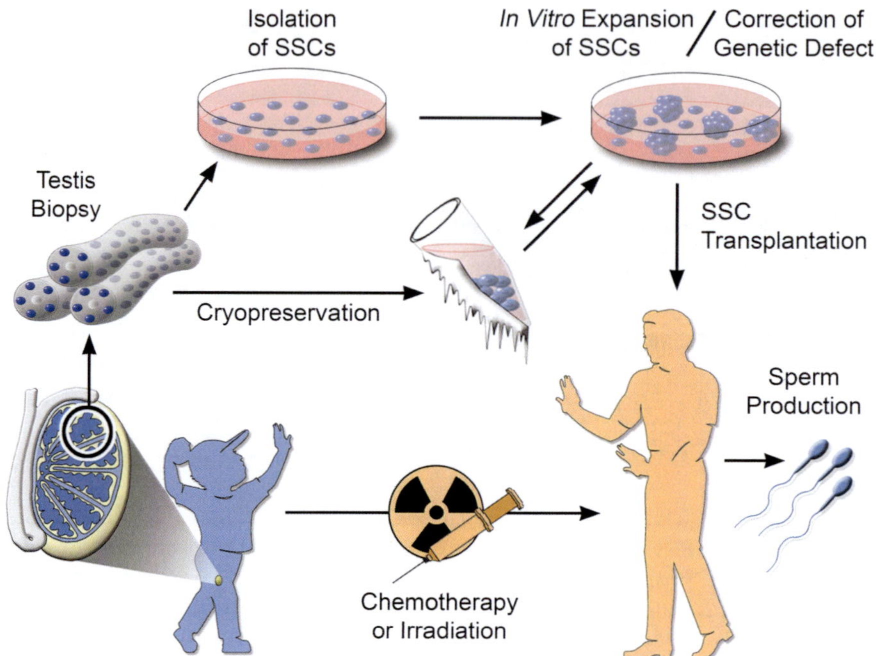

Fig. 1 Therapeutic application of SSC preservation and transplantation to restore fertility. Before undergoing cancer treatment with chemotherapy or irradiation, a testicular biopsy containing SSCs is obtained from the prepubertal patient. SSCs are isolated directly from the fresh biopsy and expanded in vitro [3]. Cultured SSCs or a biopsy containing SSCs can be cryopreserved for preserving the germline. After successful cancer treatment, SSCs are transplanted back to the patient's testis to restore fertility. The in vitro culture of SSCs also provides an opportunity for elimination of potentially contaminating cancer cells and perhaps gene therapy [4]

To achieve this, SSCs must first be isolated and frozen safely and then recovered intact after an extended period. Ultimately, the SSCs must be reintroduced into the testis (*see* Fig. 1) to regenerate spermatogenesis. We present here methodologies for each of these steps in the extensively studied murine model system. In the future, manipulation of recovered and cultured SSCs may also allow opportunities for gene therapy, including correction of genetic defects in the male germline using genome editing techniques such as the CRISPR/Cas9 system [5]. There remain substantial challenges in realizing this therapeutic vision as it would also require expanding SSC number in a long-term culture system that has yet to be developed for humans. Long-term culture to produce the number of SSCs required for successful transplantation also raises concerns about genetic and epigenetic integrity. Nevertheless, these procedures developed in rodents provide a strong foundation for preserving and modifying the germline of all species, including humans [4].

2 Materials

2.1 Isolation, Cryopreservation, and Preparation of Donor Cells

2.1.1 Donor Animals

The selection of the donor animal will depend on the strain and the genotype dictated by the goal of the experiment but will typically include a transgenic marker such as green fluorescent protein (GFP) or β-galactosidase (LacZ) to visualize spermatogenic colonies. To obtain primary Thy1+ germ cells for transplantation or developing Thy1+-derived germ cell cultures, neonatal mice 5–8 days old are used, since they have a highly enriched population of undifferentiated spermatogonia in the testis.

2.1.2 Reagents

1. Trypsin-EDTA: 0.25% Trypsin with 1 mM EDTA.
2. Hanks' balanced salt solution (HBSS) without calcium or magnesium.
3. 1 mg/mL Collagenase type IV solution prepared fresh in HBSS and sterilized by filtration.
4. 7 mg/mL DNase I stock solution prepared in HBSS and sterilized by filtration.
5. Fetal bovine serum (FBS), treated at 55 °C for 45 min. Aliquots are stored at −20 °C.
6. Dulbecco's phosphate-buffered saline supplemented with 1% FBS (DPBS-1F): DPBS, 10 mM HEPES, 1 mg/mL glucose, 1 mM pyruvate, 50 units/mL penicillin, and 50 μg/mL streptomycin.
7. Percoll solution: 30% (vol/vol) Percoll in DPBS-1F supplemented with 50 units/mL penicillin and 50 mg/mL streptomycin. Percoll solution is sterilized by filtration using a 0.22 μm membrane filter and stored at 4 °C [6].
8. CD90.2 (Thy1) microbeads.
9. Mouse serum-free medium (mSFM): Minimum essential medium-α medium (Invitrogen) with 0.2% bovine serum albumin fraction V (BSA; each batch needs to be evaluated, *see* ref. 6), 5 μg/mL insulin, 10 μg/mL iron-saturated transferrin, 7.6 μEq/L free fatty acid mixture (*see* below), 3×10^{-8} M H_2SeO_3, 50 μM 2-mercaptoethanol, 10 mM HEPES, 60 μM putrescine, 2 mM glutamine, 50 units/mL penicillin, and 50 μg/mL streptomycin [6].
10. Free fatty acid mixture: 31.0 mM Palmitic acid, 2.8 mM palmitoleic acid, 11.6 mM stearic acid, 13.4 mM oleic acid, 35.6 mM linoleic acid, and 5.6 mM linolenic acid. These are prepared as 100 mEq/L stock solution aliquots in ethanol and can be stored at −20 °C [6].
11. Human recombinant glial derived neurotropic factor (GDNF).

12. Rat recombinant GDNF Family Receptor Alpha 1/Fc chimera (GFRα1).
13. Human recombinant basic fibroblast growth factor (bFGF).
14. Dimethyl sulfoxide, tissue culture grade (DMSO).
15. Liquid nitrogen storage vessel.

2.1.3 Equipment

1. Water bath set to 37 °C.
2. 35 mm Tissue culture dish.
3. Dissection instruments.
4. Cell culture incubator set to 37 °C.
5. 15 mL Polypropylene tubes.
6. Refrigerated centrifuge.
7. 40 μm pore nylon cell strainer.
8. 50 mL Polypropylene tube.
9. 12-Well tissue culture plate, gelatinized and seeded at a density of 5×10^4 cells/cm^2 with STO feeders (SNL76/7 from ATCC) that have been treated with 10 μg/mL mitomycin C [6].
10. Trypan blue.
11. Neubauer hemocytometer.
12. MS columns.
13. MACS Separator with MultiStand.
14. 1 mL Cryotubes.
15. Nalgene freezing container filled with isopropanol.
16. −80 °C Freezer.

2.2 Germ Cell Transplantation into Recipient Mice

2.2.1 Recipient Animals

Recipient mice should be devoid of endogenous spermatogenesis to avoid competition between recipient and donor spermatogonia. Mutant mice lacking endogenous germ cells, such as W mice, can be used directly as recipients [7]. Alternatively, endogenous germ cells can be removed in the recipient by Busulfan treatment described in Subheading 3.6, in which case recipients are treated with Busulfan at age 4–6 weeks. Ideally, the donor and recipient should be immunologically compatible, but immunocompromised mice (e.g., nude mice) can be used as recipients in cases where incompatibility cannot be avoided, such as xenotransplantation.

2.2.2 Reagents

1. DMSO.
2. Busulfan.
3. Analgesic.
4. Anesthetic.
5. Iodine solution.

2.2.3 Equipment

1. Glass scintillation vials, 20 mL.
2. Heating block for 50 mL tube.
3. 1 mL Syringe.
4. 26 ½ Gauge needle.
5. Weighing scale for mice.
6. Fine surgical instruments.
7. Electric clipper for removing hair.
8. Sterile surgical drape.
9. Dissecting microscope (with a magnification range of 6× to 20×).
10. V-shaped cardboard cutouts: Squares of autoclaved note card (approx. 1 cm × 1 cm with a notch cut from one edge, *see* Fig. 3).
11. Polyethylene tubing (0.034 × 0.050 × 0.008).
12. mSFM (*see* above section).
13. Trypan blue.
14. Borosilicate glass needles: Start with a polished borosilicate glass capillary, 4″/100 mm in length, with an internal diameter of 0.75 mm and an external diameter of 1 mm. Silicone coat the capillaries by putting them into a glass tube containing Sigmacote solution and invert several times to allow the silicone to flow into capillaries. Remove silicone and wash 3× with methanol. Oven-dry for 2 h. Once dry, using a pipette puller, pull the capillary into a sharp point. Carefully break the tip of the point with forceps. Attach polyethylene tubing to the needle and to a syringe filled with water. Secure the needle in a clamp and, maintaining water flow through the needle, bevel the tip with a grinder to a ~30° angle and with an outside diameter of approximately 40 μm. Once the desired bevel is achieved, dry the needle by blowing air through it.
15. 1.5 mm Microelectrode holder.
16. Microinjector.
17. Fiber-optic illuminator.
18. Silk sutures, 6-0.
19. Surgical clips, 7.5 mm.
20. Hot plate.
21. Hot bead sterilizer.

3 Methods

3.1 Preparation of Testicular Cell Suspension

Isolation and propagation of stem cells outside the body are vital for germline stem cell therapy. For medical applications the source of these SSCs could be a biopsy or cryopreserved testicular tissue. For mouse models the SSCs are derived from whole testes of newborn or adult animals (Fig. 2). Regardless of the source, the first step in isolation of SSCs is digestion of testis sample into a single-cell suspension.

1. Place 0.25% trypsin-EDTA in a 37 °C water bath to warm up. Prepare two 35 mm dishes with 2 mL ice-cold HBSS in each.
2. Collect 3–5 neonatal mice aged 5–8 days postpartum (dpp). Adult mice or adult/neonate rats can also be used (*see* **Note 1**). Sacrifice animals via institutionally approved method.

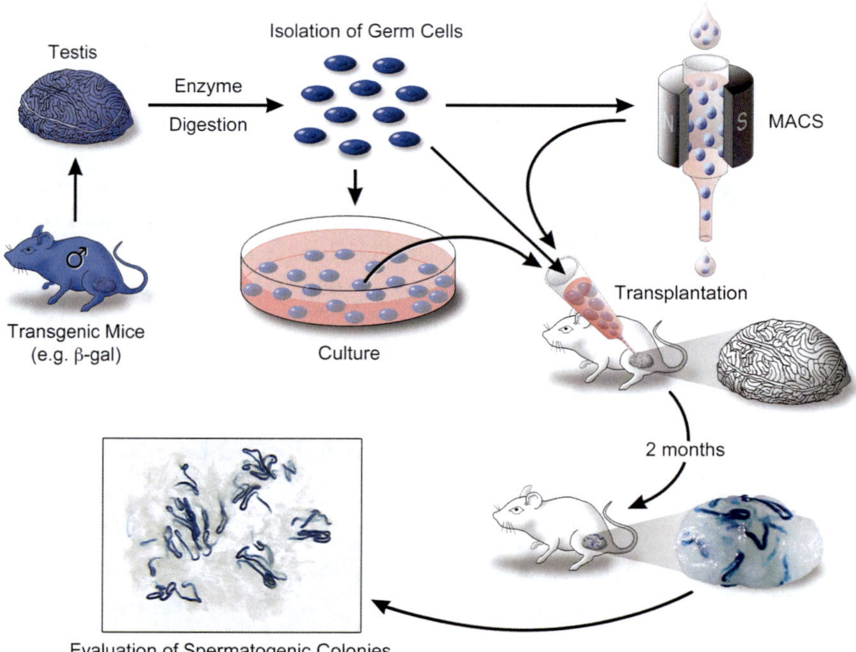

Fig. 2 Germ cell transplantation and quantitative assessment of SSC activity. The source of the donor cells for transplantation into recipient testes can be a testicular single-cell suspension, in vitro-cultured cells enriched for spermatogonial stem cells, or cells fractionated by magnetically activated cell sorting (MACS). Regardless of the source, it is desirable for cells to contain a marker gene to visualize spermatogenic colonies (e.g., β-galactosidase, which in the presence of the substrate X-gal stains blue). Eight weeks after transplantation, donor-derived spermatogenesis can be readily detected in the recipient testis as blue colonies and can be counted. Each spermatogenic colony generally develops from a single SSC [9]; thus, a count of the number of colonies provides a quantitative estimation of the number of SSCs in the donor population. The colony length demonstrates the degree of expansion (modified from [3, 10])

3. Remove testes with sterile instruments, and place into a 35 mm petri dish with ice-cold HBSS.
4. Decapsulate testes and transfer to a fresh dish with ice-cold HBSS.
5. Bring dish to tissue culture hood, aspirate off the HBSS, and place testes in 4.5 mL of 1 mg/mL collagenase with 1 mL of 7 mg/mL DNase I. Using a 5 mL serological pipette gently pipette the testes up and down several times to facilitate dissociation of the seminiferous tubules (*see* **Note 2**).
6. Place dish in 37 °C incubator for 10 min (*see* **Note 3**). Pipette again several times and transfer to a 15 mL polypropylene conical centrifuge tube.
7. Let the 15 mL tube to stand for 1–2 min to allow the tubules to settle at the bottom. Remove the supernatant, add 10 mL HBSS, invert the tube several times, and spin again. This step should be repeated twice to wash the tubules.
8. Resuspend in 4.5 mL of trypsin-EDTA with 0.5 mL of 7 mg/mL DNase I solution. Place in 35 mm culture dish and place in incubator. Check after 5 min and pipette with a p1000 several times to break up remaining clumps. Incubate for an additional 3 min at 37 °C (*see* **Note 4**).
9. Add FBS to 10% of final volume to stop the digestion. Add 0.5 mL DNase I solution and pipette with a p1000 several times (*see* **Note 5**).
10. Assemble a 40 μM nylon cell strainer over a sterile 50 mL tube. Add 1 mL HBSS to the cell strainer and discard. Filter the cell suspension through the strainer and wash strainer with 5–10 mL HBSS.
11. Centrifuge the cell suspension at 600 × *g* at 4 °C for 7 min.
12. Wash the cells in HBSS and centrifuge again.
13. Resuspend in mSFM and count cells with a hemocytometer (*see* **Note 6**).

3.2 Isolation of Primary Thy1+ Germ Cells Using MACS

While SSCs do not have a single identifying cell surface marker, considerable enrichment can be achieved by selecting for Thy1, a marker displayed on undifferentiated spermatogonia in the mouse. Magnetically activated cell sorting (MACS) using anti-Thy1 antibody provides a convenient method of enriching murine SSCs (*see* **Note 7**).

1. Follow steps in Subheading 3.1 (above) to obtain testicular single-cell suspension from adult or neonate testes.
2. Discard the supernatant and resuspend the cell pellet in 5 mL of DPBS-1F. Perform cell count and adjust the volume to retain ~2×10^7 cells per 5 mL of DPBS-1F.

3. Gently overlay 5 mL of cell suspension into a 15 mL tube containing 2 mL of ice-cold 30% Percoll. Care must be taken not to disturb the interphase between Percoll and cell suspension. Spin the tube at 600 × *g* for 7 min at 4 °C.
4. Discard the supernatant carefully by first pipetting out the trapped cellular debris at the interphase followed by removing the entire supernatant. Resuspend the pellet in ice-cold DPBS-1F and perform cell count.
5. Spin the tube at 600 × *g* for 7 min at 4 °C to wash the cell pellet. Discard the supernatant and resuspend the pellet in 90 μL of DPBS-1F when the cell number is 6 × 10^6 cells or less. Use 180 μL of DPBS-1F if more than 6 × 10^6 cells.
6. Add 10 μL or 20 μL of CD90.2 (Thy1) magnetic microbeads to 90 μL or 180 μL of cell suspension, respectively, and gently tap to mix it. Incubate the cells by placing the tube on ice for 20 min and tap the tube at an interval of 10 min.
7. At the end of incubation add 2 mL of DPBS-1F to the cell suspension to dilute it and spin it at 600 × *g* for 7 min at 4 °C. While the cells are spinning assemble the MACS separation column and the magnetic separation unit. Wash the column once with 0.5 mL of DPBS-1F.
8. Discard the supernatant at the end of the spin and resuspend the cell pellet in 1 mL of DPBS-1F. Make sure to disperse the clumps by gently pipetting; otherwise the clumps will clog the column during subsequent washing step. Dispense 1 mL of cell suspension onto the previously equilibrated MACS column and allow it to flow through by gravity.
9. Wash the column three times, each with 0.5 mL of DPBS-1F. After the final wash, transfer the column from the scaffold and place it on a 15 mL tube. Add 1 mL of DPBS-1F to the column and place the plunger. Apply constant positive pressure to elute the magnetically bound cells (*see* **Note 8**).
10. Spin the final eluate at 600 × *g* for 7 min at 4 °C. Discard the supernatant, resuspend the pellet in 1 mL of mouse serum-free medium (without growth factors), and perform cell count. Wash the cells twice with mSFM.
11. At the end of final wash perform a cell count. Adjust the cell concentration to desired density by adding appropriate volume of mouse serum-free medium containing growth factors at the following proportion: 20 ng/mL of GDNF, 150 ng/mL of GFRα1, and 1 ng/mL of bFGF.
12. Typical recovery of Thy1+ germ cells using MACS separation is within the range of 1.6 × 10^5 to 2.2 × 10^5 cells starting with 4–8 testes (2–4 pups). Given the above guidelines, the isolation of Thy1+ germ cells can be scaled up or down.

3.3 Preparation of Donor Cells from Cultured Thy1+ Cells

Development of in vitro culture techniques for mouse and rat SSCs has provided powerful tools for the understanding of their biology and mechanisms of regulation. Ultimately, cells derived from SSC culture have the potential to be directly reintroduced into a recipient animal. In the human, the SSC population would be reintroduced into the testis following elimination of cancer cells in the patient and in the donor cell population. This section assumes that you start with an established Thy1+ mouse SSC culture on a STO feeder layer as described in [6].

1. Determine the number of cultured Thy1+ cells required at the beginning of the experiment depending on the number of testis to be transplanted and the experimental question to be addressed.
2. Seed germ cells derived from established SSC culture at ~1×10^5 cells per well of a 12-well plate at least 4–5 days before the day of experiment. Change the media every 2 days but do not allow the cells to overgrow as that will allow spontaneous differentiation of the undifferentiated spermatogonia in the cultured germ cell clumps.
3. On the day of the transplantation experiment, remove the culture from the incubator and add 1.5 mL of pre-warmed mSFM to each well. Gently blow the cells off by pipetting the medium in the well [11]. Take care not to rupture the feeder cell layer. Collect and pool the blown-off cells in a 15 mL tube and spin at $600 \times g$ for 5 min at 4 °C. Discard the supernatant and resuspend the cell pellet in 1 mL of pre-warmed 0.25% trypsin-EDTA solution. Add 50 μL of DNaseI solution, gently mix, and incubate the tube in a water bath at 37 °C for 2 min.
4. Stop trypsinization by adding FBS to 10% of final volume and 50 μL of DNase I solution. Spin the tube at $600 \times g$ for 7 min at 4 °C, discard the supernatant, resuspend the cell pellet in 1 mL of mSFM, and count. The resulting germ cell suspension obtained by this method is ~96% pure [11]. The final volume is adjusted based on the number of cells to be used for transplantation.

3.4 Cryopreservation of Testis Cells

The ability to cryopreserve testicular tissue or cells and recover germ cells later is required for the restoration of spermatogenesis in the patient's testes following elimination of cancer cells. Here we outline a cryopreservation technique that has been successful in restoring spermatogenesis in the rodent model. This technique is a foundation for cryopreservation of testicular cells and SSCs in other species [12].

1. Prepare 2× freezing media as follows: mix DMSO, mSFM (without growth factors), and FBS in the ratio 1:2:2 in a

15 mL tube. It is important to make this media fresh before freezing. Place the mixture on ice for at least 5 min to chill.

2. Resuspend cells from a fresh testis isolation (adult/neonate) or from culture in mSFM (containing 40 ng/mL GDNF, 300 ng/mL GFRα1, and 2 ng/mL bFGF) at a cell density of 1–4 × 10^7 cells/mL in a 50 mL tube.
3. Add an equal volume of the 2× freezing media dropwise to the suspended cells while swirling the tube for a final concentration of 0.5–2 × 10^7 cells/mL.
4. Very gently aliquot 1 mL into each cryotube and place the tubes in the Nalgene freezing container. The cells are fragile and should be pipetted with care.
5. Place into −80 °C freezer overnight.
6. Move cells to liquid nitrogen storage the next day.

3.5 Thawing of Frozen Testis Cells

Current cryopreservation and thawing techniques for SSCs are based on principles developed for somatic cells because, unlike spermatozoa, SSCs resemble somatic cells in structure [13]. Human and baboon cells have been successfully cryopreserved and transplanted into mice, where the cells colonized the basement membrane but did not undergo further differentiation [12, 14]. Mouse SSCs cryopreserved for approximately 12 years when thawed could be cultured and transplanted and colonize recipient testes resulting in spermatogenesis and normal progeny [12]. These studies established a foundation for restoration of fertility in prepubertal boys treated for cancer.

1. Remove cryotube from liquid nitrogen and place directly into a 34–37 °C water bath. Swirl the tube until thawed (~2 min). Remove from water bath and spray vial with 70% ethanol before bringing into tissue culture hood.
2. Transfer contents of cryotube into a 50 mL tube. Very gently add mSFM dropwise to the thawed cells to a total volume of 9 mL while swirling the tube. Use 1 mL of mSFM to rinse out the cryotube and add it to the 50 mL tube bringing the volume up to 10 mL. The cells are fragile, so pipette slowly and avoid excessive shear forces.
3. Transfer the suspension to a 15 mL conical tube.
4. Spin cells at 600 × *g* at 4 °C for 7 min. Remove supernatant and resuspend in 5 mL mSFM. Repeat this step.
5. Count cells in a hemocytometer and check cell viability via trypan blue exclusion. For mouse pups, cell viability can be expected to be 68.8% ± 1.31 (SEM) [12], although for older mice and for rats the viability may be lower. If the cell viability is low, dead cells can be removed by 30% Percoll fractionation (*see* Subheading 3.2).

6. Spin cells at 600 × *g* for 7 min. Remove supernatant and resuspend in mSFM (containing growth factors at the following proportions: 20 ng/mL of GDNF, 150 ng/mL of GFRα1, and 1 ng/mL of bFGF) at the desired cell concentration.

3.6 Busulfan Injection into Recipient Mice

To prepare recipient males to study SSCs in the mouse, the cytotoxic agent Busulfan is often used to remove endogenous germ cells from the seminiferous tubules. Busulfan treatment mimics the effect of anticancer chemotherapeutic cytotoxic agents that results in a loss of endogenous germ cells in prepubertal boys. The techniques described below are applicable with modification to other species.

1. In a fume hood, prepare a fresh solution by dissolving Busulfan in DMSO in a glass scintillation vial to obtain a concentration of 8 mg/mL. Dilute the resulting solution by slowly adding an equal volume of warm distilled water (~56 °C) to obtain 4 mg/mL final concentration. To prevent Busulfan from precipitating add water slowly along the wall of the glass vial so that the water overlays the DMSO solution. Maintain these two phases until ready to inject. During the entire procedure from preparation to injection maintain the temperature of the Busulfan solution between 35 °C and 40 °C, for example in a heating block.
2. Just before injection, thoroughly mix the DMSO and water phases by gently swirling or inverting the vial several times. Draw the solution slowly into a 1 mL syringe fitted with 26½ gauge needle ensuring that no bubbles form inside the syringe barrel.
3. Weigh each mouse to determine body weight. Inject Busulfan into the male mouse intraperitoneally at a dose of 60 mg/kg body weight for F_1 hybrid recipients (e.g., B6129SF1/J) aged between 4 and 6 weeks. The exact dose varies depending on the strain of mice (*see* **Note 9**).
4. Male mice treated with Busulfan are maintained for ~6 weeks before they are used for transplantation. The timing window is critical to allow depletion of all endogenous germ cells within the seminiferous tubules. Efficient elimination of endogenous germ cells facilitates the migration of donor spermatogonial stem cells to colonize the vacant niches and give rise to donor-derived spermatogenic colonies.

3.7 Microinjection of Donor Germ Cell into Recipient Mice

1. Resuspend cells (from Subheadings 3.1–3.3 and 3.5) for transplantation at the desired concentration, up to 10^8 cells/mL. Microinjection is critically dependent on the quality of the prepared cells, especially at high cell densities (*see* **Note 10**). Prepare ~25 μL of cell suspension for each recipient animal, of

which a final volume of 5–10 μL of cell suspension is injected per testis.

2. Bring sterile instruments, analgesic, and anesthetic into the surgical room.
3. Select the appropriate mouse strain for the recipient (*see* Subheading 2.1.1).
4. Weigh each mouse. Administer appropriate analgesic for pain management.
5. Administer anesthetic by intraperitoneal injection according to institutional guidelines.
6. Remove hair at the surgical site on the abdomen with clippers and transfer the mouse to the surgical table.
7. Wipe the surgical site with iodine solution. Pinch mouse toe and check for palpebral reflex to confirm proper surgical anesthesia level.
8. Place sterile drape over mouse and make an approximately 1 cm midline incision through skin and abdominal wall.
9. Transfer mouse to the microscope platform. Locate the first testis and using forceps to hold the fat pad draw it out through the abdominal opening onto a sterile V-shaped cardboard that rests on the drape. This helps to secure the position of the testis (*see* Fig. 3f).
10. Cut ~2 cm of polyethylene tubing. Fill the tubing with 5–10 μL volume of cell suspension with a p20 micropipette (Fig. 3a). Cells prepared from previous sections should be resuspended in mSFM with trypan blue added at a 0.04% final concentration to visualize the suspension (Fig. 3b).
11. Attach one end of the tubing to the needle drawn from the borosilicate glass capillary (Fig. 3c).
12. The microelectrode holder is attached to 1 mL syringe (cut to receive the tubing) and this in turn is affixed to the metal micropipette holder of Eppendorf microinjector (Fig. 3d).
13. Insert the tubing entirely into the 1.5 mm microelectrode holder so that the needle can be held in place by the orange tightening ring (Fig. 3e).
14. Locate the efferent duct and insert the tip of the needle a short length into the duct toward the rete of the testis (Fig. 3h). Inject the cells into the testis via the rete into the seminiferous tubules (*see* **Note 11**).
15. Watch the spread of the dye through the seminiferous tubules over the surface area of the entire testis. A good injection will result in 75–85% of the surface tubules filled with the dye (*see* **Note 12**).

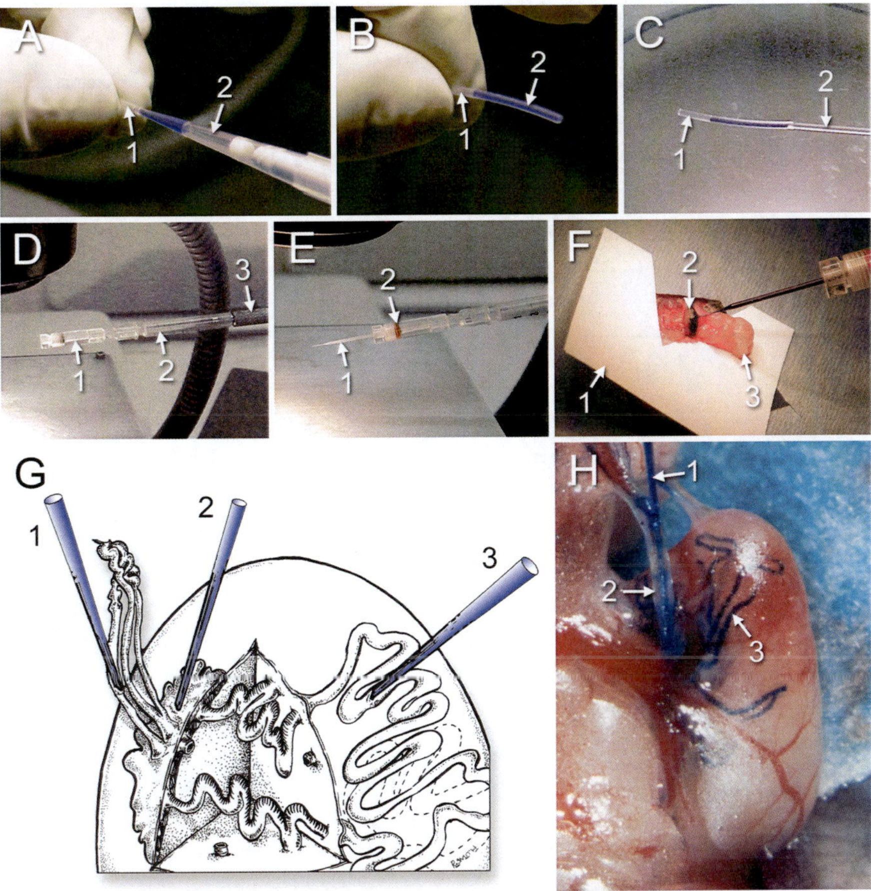

Fig. 3 Microinjection of donor germ cells into the recipient testis: (**a**) Loading the polyethylene tubing (1) using a p20 pipette (2). (**b**) The polyethylene tubing (1) containing the blue cell suspension (2). (**c**) Polyethylene tubing (1) is attached to the needle (2). (**d**) The microinjector scaffold, consisting of the microelectrode holder (1), the syringe (2), and Eppendorf micropipette holder (3). (**e**) The needle is attached to the injection assembly. The tubing is inserted into the micropipette holder and the needle (1) is secured in place using the orange tightening ring (2). (**f**) The V-shaped sterile cardboard (1) is used to stabilize the recipient testis, (2), which is removed from the body cavity by the fat pad (3). (**g**) Cartoon showing the possible entry points for the needle: SSCs can be injected into an efferent duct (1), into the rete testis (2), or directly into a seminiferous tubule (3). (**h**) Injection of donor cells via the efferent duct. The needle (1) is inserted into one of the efferent ducts (2) and its position is secured. Donor cells mixed with loading dye flow from the needle into the efferent duct, then to the rete testis, and finally into the seminiferous tubules (3). A successful injection leads to filling up of most, if not all, of the seminiferous tubules on the surface (**g** and **h** modified from [8])

16. Place the testis back into the abdominal cavity and repeat the procedure with the other testis.
17. Suture the muscle layer with 6-0 silk thread using interrupted ligatures. Close the skin with 7.5 mm surgical clips.
18. Transfer mice into a cage and place the cage on warm plate at 35–37 °C.

19. Monitor mice every few minutes until they recover fully from the effect of anesthesia.
20. Clean the instruments and sterilize in a hot bead sterilizer between surgeries.
21. Transplanted mice should be monitored on a warm plate for 3–7 days postsurgery for signs of pain and distress. The surgical site should be checked daily during the period for signs of continued closure, inflammation, or infection.

4 Notes

1. Adult mice or neonatal/adult rats can be used for making testis whole-cell suspension. These protocols have been written with mouse cells in mind but are also suitable for rat cells. Rat serum-free media, as detailed in [6], should be substituted for mSFM for all procedures in this chapter involving rat cells.
2. Collagenase digestion is not required for neonatal mouse testes but is recommended for adult mouse testes or for rat testes. This volume can be scaled up for larger amounts of testicular tissue.
3. Collagenase and trypsin digests can be carried out in a centrifuge tube placed in a 37 °C water bath instead, but the tube should be agitated periodically by flicking the tube to prevent the tissue collecting at the bottom.
4. This step assumes testes from three to five neonatal mice and this volume can be scaled up or down as necessary, maintaining the 9:1 ratio of trypsin-EDTA to DNase I as a starting point. Additional DNase I can be added if the solution remains very viscous or visible clumps persist after digestion. For the rat, a 4:1 ratio is appropriate and two adult rat testes can be digested in 20 mL trypsin-EDTA and 5 mL DNase I.
5. If the cells are still clumped or the solution is viscous, add another 0.5 mL DNaseI solution and pipette again. This step can be repeated as necessary to remove DNA in solution and prevent aggregation.
6. Typical cell recovery from five to eight dpp mice is $1–2 \times 10^6$ cells/testis. For 9–10 dpp rat pups typical recovery is $10–13 \times 10^6$ cells/testis.
7. Thy1 does not enrich rat SSCs as efficiently as mouse SSCs. EpCAM is a superior marker for rat SSCs. Therefore in order to select for rat SSCs, anti-EpCAM antibody should be used in place of anti-Thy1 for rat isolations as described in [6].
8. If a highly enriched population of Thy1+ cells is required, the eluate can be passed through another round of separation by

Table 1
Busulfan dosage for common strains of mice

Mouse strain	Busulfan dose (mg/kg)
C57BL/6J	44
B6129SF1/J	60–65
C57BL/6J × FVB	50
B6D2F1	50–55
Nudes (J/Nu)	40–44

following **steps 12–14**. Alternatively, to maximize Thy1+ recovery at the expense of purity, the flow-through can be passed through a fresh column, repeating **steps 12–14**.

9. The tolerance for Busulfan varies between mouse strains and with age. Refer to Table 1 for guidance, but final dosage should be determined empirically, especially for nudes or other strains with less tolerance.
10. It is important to prepare a clean cell suspension for injection in order to minimize clogging of the needle tip with dead or clumped cells. To avoid this, it is recommended that the cell suspension should be filtered using a 40 μm cell strainer (Subheading 3.1) or fractionated using 30% Percoll (Subheading 3.2) or both.
11. Maintain the cell suspension at 4 °C throughout the procedure, except during loading and injecting into the efferent duct.
12. Improper needle positioning can lead to filling of the interstitial spaces between the seminiferous tubules. Avoid overfilling with a higher volume or concentration of cells than recommended, since it can lead to buildup of pressure inside the testis. Both interstitial leakage and overfilling may result in hardening of the testis. This can be mitigated by making a small puncture in the tunica to release the extra pressure or to drain out the interstitial accumulation.

Acknowledgments

We thank James Hayden, RBP, FBCA, for assistance in preparation of figures and Mary Avarbock for advice and expert assistance. This work was supported by the National Institute of Child Health and Human Development (R.L.B.) and the Robert J. Kleberg, Jr. and Helen C. Kleberg Foundation (R.L.B.).

References

1. Katz DJ, Kolon TF, Feldman DR, Mulhall JP (2013) Fertility preservation strategies for male patients with cancer. Nat Rev Urol 10 (8):463–472
2. Tournaye H, Dohle GR, Barratt CL (2014) Fertility preservation in men with cancer. Lancet 384(9950):1295–1301
3. Kubota H, Brinster RL (2006) Technology insight: in vitro culture of spermatogonial stem cells and their potential therapeutic uses. Nat Clin Pract Endocrinol Metab 2:99–108
4. Brinster RL (2007) Male germline stem cells: from mice to men. Science 316:404–405
5. Sato T, Sakuma T, Yokonishi T, Katagiri K, Kamimura S, Ogonuki N, Ogura A, Yamamoto T, Ogawa T (2015) Genome editing in mouse spermatogonial stem cell lines using TALEN and double-nicking CRISPR/Cas9. Stem Cell Rep 5(1):75–82
6. Kubota H, Brinster RL (2008) Culture of rodent spermatogonial stem cells, male germline stem cells of the postnatal animal. Methods Cell Biol 86:59–84
7. Ogawa T, Dobrinski I, Avarbock MR, Brinster RL (2000) Transplantation of male germ line stem cells restores fertility in infertile mice. Nat Med 6:29–34
8. Ogawa T, Aréchaga JM, Avarbock MR, Brinster RL (1997) Transplantation of testis germinal cells into mouse seminiferous tubules. Int J Dev Biol 41(1):111–122
9. Kanatsu-Shinohara M, Inoue K, Miki H, Ogonuki N, Takehashi M, Morimoto T, Ogura A, Shinohara T (2006) Clonal origin of germ cell colonies after spermatogonial transplantation in mice. Biol Reprod 75 (1):68–74
10. Brinster RL (2002) Germline stem cell transplantation and transgenesis. Science 296 (5576):2174–2176
11. Oatley JM, Avarbock MR, Telaranta AI, Fearon DT, Brinster RL (2006) Identifying genes important for spermatogonial stem cell self-renewal and survival. Proc Natl Acad Sci U S A 103(25):9524–9529
12. Wu X, Goodyear SM, Abramowitz LK, Bartolomei MS, Tobias JW, Avarbock MR, Brinster RL (2012) Fertile offspring derived from mouse spermatogonial stem cells cryopreserved for more than 14 years. Hum Reprod 27(5):1249–1259
13. Avarbock MR, Brinster CJ, Brinster RL (1996) Reconstitution of spermatogenesis from frozen spermatogonial stem cells. Nat Med 2 (6):693–696
14. Nagano M, Patrizio P, Brinster RL (2002) Long-term survival of human spermatogonial stem cells in mouse testes. Fertil Steril 78 (6):1225–1233

Chapter 15

Human-Monkey Chimeras for Modeling Human Disease: Opportunities and Challenges

Alejandro De Los Angeles, Insoo Hyun, Stephen R. Latham, John D. Elsworth, and D. Eugene Redmond Jr.

Abstract

The search for a better animal model to simulate human disease has been a "holy grail" of biomedical research for decades. Recent identification of different types of pluripotent stem cells (PS cells) and advances in chimera research might soon permit the generation of interspecies chimeras from closely related species, such as those between humans and other primates. Here, we suggest that the creation of human-primate chimeras—specifically, the transfer of human stem cells into (non-ape) primate hosts—could surpass the limitations of current monkey models of neurological and psychiatric disease, but would also raise important ethical considerations concerning the use of monkeys in invasive research. Questions regarding the scientific value and ethical concerns raised by the prospect of human-monkey chimeras are more urgent in light of recent advances in PS cell research and attempts to generate interspecies chimeras between humans and animals. While some jurisdictions prohibit the introduction of human PS cells into monkey preimplantation embryos, other jurisdictions may permit and even encourage such experiments. Therefore, it is useful to consider blastocyst complementation experiments more closely in light of advances that could make these chimeras possible and to consider the ethical and political issues that are raised.

Key words Interspecies chimeras, Pluripotency, Chimeras, Reprogramming, Naïve pluripotency, Naïve pluripotent stem cells, Primed pluripotent stem cells, Nonhuman primates, Human-monkey chimeras, Disease modeling, Stem cells

1 New Approaches to an Old Problem

Neurological and psychiatric diseases are a devastating problem, causing profound human suffering and disease burden worldwide. The World Health Organization estimates that at least one billion people are affected by neurological disease; the number is expected to increase considerably in the future [1]. Neurological disorders are also a major cause of mortality and comprise 12% of total deaths globally. Currently available treatment options are often fruitless. Failure rates for experimental central nervous system drugs are higher compared with other classes of drugs [2]. One possible

Insoo Hyun and Alejandro De Los Angeles (eds.), *Chimera Research: Methods and Protocols*, Methods in Molecular Biology, vol. 2005, https://doi.org/10.1007/978-1-4939-9524-0_15, © Springer Science+Business Media, LLC, part of Springer Nature 2019

roadblock is the inadequate quality of existing animal models, which impedes elucidation of disease mechanisms and development of new treatments.

One strategy for modeling of neurological disease is to generate disease-relevant cell types from patient-specific induced pluripotent stem (iPS) cells [3–6]. The defining features of human iPS cells—indefinite self-renewal in culture and capacity to differentiate into any cell type—in principle allow for access to an endless supply of disease-relevant cells. Further, iPS cell-derived cells are genetically identical to the source patients. When combined with advances in generating neural cell types by directed differentiation of hiPS cells, researchers have probed disease-specific effects on a relevant cell type. This experimental strategy has been successfully applied to establish correlations between patient-specific genetic mutations and abnormal neuronal phenotypes for a range of highly penetrant neurological disease [7, 8].

Classical strategies for modeling neurological disease using patient-specific iPS cells involve differentiation in a dish. However, in vitro differentiation possesses critical limitations, including the failure to achieve functional maturation of in vitro-generated human PS cell derivatives that tend to exhibit immature, fetal-like features. Moreover, current methodologies are not compatible with the production of complex three-dimensional tissues, preventing scientists from assessing connectivity and systems-level functionality of disease-specific cells. Consequently, new approaches are required to generate more sophisticated hiPS cell-based models of neurological disease.

The limitations of in vitro disease modeling underscore a need to return to the in vivo approaches. For example, it is now possible to establish mice within which a significant percentage of host glia are patient derived [9]. Indeed, a recent study generated human iPS cell mouse chimeras using glial progenitor cells derived from iPS cells from patients with childhood-onset schizophrenia [10]. These mice were reported to exhibit unusual behaviors, which included increased anxiety, antisocial traits, and perturbed sleep. A similar approach has been applied to modeling Huntington's disease [11]. However, one challenge for modeling neurological and psychiatric conditions using rodents as a host animal is knowing whether the model faithfully recapitulates signs of the disease. The use of nonhuman primates (NHPs) as a host animal may generate more interpretable data.

Because of their similarities to humans, NHPs are essential models for studying neurological disease. Nonetheless, there remain neurological characteristics unique to humans. Knowledge of relevant convergent and divergent features must be inextricably linked to the choice of approach for studying neurological and psychiatric disease. While NHP models for neurological disease have existed for decades, rapid advances in CRISPR/Cas9-

mediated genome editing are altering how we model disease using NHPs [12–14]. For example, proof-of-principle experiments applying CRISPR/Cas9 to early primate embryos have generated knockout monkeys for the PPARG and RAG1 loci [15].

Nonetheless, there exist three hurdles associated with the use of CRISPR/Cas9-mediated genome editing. First, while CRISPR/Cas9 accurately cleaves its target genomic loci, it also possesses off-target activity that can induce undesired genetic changes. Second, off-target or delayed Cas9 activity can also cause genetic mosaicism, where an animal is composed of cells with different genotypes. A third technical challenge is creating NHPs with mutations at multiple loci. Overcoming this obstacle may be achieved by unifying genome editing with somatic cell nuclear transfer (SCNT)-based cloning of monkeys, which will in principle enable generation of transgenic NHPs by using donor nuclei from cultured gene-edited cells [16].

Given the limitations posed by CRISPR/Cas9 approaches, we believe that new advances in human iPS cell research may provide a fruitful alternative method for creating appropriate NHP models of neurological diseases. It is now broadly accepted that in vitro pluripotency manifests as a continuum of different cellular states [17]. At one polar extreme is the naïve state, which reflects unrestricted cellular potency. At the other end is the primed state, where cells are poised for differentiation [18]. The key distinction between naïve and primed PS cells is the ability to generate chimeras—the capacity to contribute to all three germ layers when injected into a preimplantation embryo [18, 19]. While rodent PS cells are in a naïve state because they can form chimeras, conventional PS cells in primates and human are likely to reside in a non-naïve pluripotent state as primate PS cells cannot form chimeras when introduced into preimplantation embryos [20].

In recent years, interest has grown in understanding the xenogenicity of PS cells—the capacity of PS cells from one species to contribute to the embryos of another species [21]. Rodent PS cells are xenogeneic as one can reproducibly generate interspecies chimeras between mice and rats [22]. To our knowledge, the introduction of human PS cells into embryos of other species has not given rise to live interspecies chimeras [21, 23]. It has been suggested that matching developmental stage of donor cells and host embryos is essential for chimera formation [24]. To match donor cell stage with host embryos, recent studies have attempted to reset the developmental stage of human PS cells toward a naïve state with the aim of conferring chimera competency. While the precise constellation of molecular and biological characteristics that identify a naïve or primed PS cell is contentious, naïve PS cells—at least in rodents—possess a unique ability to form chimeras. Therefore, a discussion of human naïve and primed pluripotency is germane because the availability of chimera-competent human PS cells is an essential reagent for generating interspecies chimeras. However, attempts to generate interspecies chimeras using human naïve-like cells and either mouse or pig host embryos have produced very low rates of chimerism [21, 23]. These initial results suggest that interspecies chimerism with naïve-like human PS cells is limited, which may reflect species barriers beyond matching developmental timing. Nonetheless, while modest, the existence of any human-pig cross-species chimerism provides hope that it may be possible to achieve significant levels of human chimerism in large animals [21].

These data provoke questions regarding why the levels of chimerism observed in these experiments are very low. One interpretation is that the developmental stage of donor cells and host embryos has not been sufficiently matched [24–26]. This could be because naïve pluripotency has not been appropriately instated in donor cells [17, 27]. In contrast, host blastocysts at the time of embryo injection may not be developmentally synchronized with putative donor naïve cells [28]. In this regard, the nature of chimera competency in primates may fundamentally differ from the rodent paradigm, such that currently defined "naivete" does not correspond to chimera-forming ability. Preliminary evidence suggests that "intermediate" types of PS cells, rather than naïve-like types, apparently possess a higher capacity to chimerize embryos of other species [21, 29]. Finally, the evolutionary distance between donor and host species may play a role [30]. The existence of rat-mouse chimeras suggests that the results obtained when introducing human PS cells into mouse and pig host embryos will differ if host embryos from more closely related species such as NHPs were used [21, 22]. To develop effective strategies to lower species barriers, it may be instructive to study chimerism in early-stage human-monkey embryos cultured to postimplantation stages to identify barriers to human-non-primate interspecies chimera formation [30].

In short, transformative advances in stem cell and chimera research necessitate revisiting the questions surrounding human-monkey chimeras.

2 Considering Human-Monkey Chimeras: Modeling Neurological and Psychiatric Disease

Despite the advantages offered by genetically engineered NHPs, current NHP models for neurodegenerative diseases are so limited as to require consideration of the benefits of human chimeras. Below, we describe certain deficiencies of different NHP models where production of NHPs with high-grade human chimerism could allow more faithful modeling of human disease (Table 1). In Table 1, we have provided a side-by-side comparison of the

Table 1
Approaches for modeling neurological and psychiatric disease using NHPs

Model type	Aging	Injury/toxin	Transgenic	Chimera
Example	• Parkinson's disease (PD) • Alzheimer's disease (AD)	• MPTP model of Parkinson's disease • 3-NP model of Huntington's disease • PCP model of schizophrenia	• Transgenic Huntington's disease model	N/A
Pros	• Can serve as partial model for neurodegenerative disease	• Can recapitulate most, but not all disease hallmarks	• Study function of disease-associated genes and non-coding regions in genetically defined system • For monogenic diseases, may mimic pathology more accurately than chemical lesions	• Study patient cells in human-like setting for modeling disease, drug screening • Modeling of polygenic diseases • Study human-specific features • Model inherited diseases for which causative mutations are unknown
Cons	• Aging is not the same as neurodegenerative disease • Accordingly, differences from corresponding human neurodegenerative disease (e.g., for modeling AD, cognitive decline with age, but neuronal loss lacks similarity with that observed in AD)	• Does not recapitulate all disease hallmarks, particularly mechanism of cell death • Monkey-to-monkey variation • Does not replicate human condition as faithfully as genetic models	• Difficult to model polygenic disease, multiple risk alleles of small effect • Off-target effects of CRISPR/Cas9 • Mosaicism from delayed and/or multiple cleavage events from Cas9 injection into early embryos • Cannot model inherited disease where causative mutation unknown	• Difficult to control chimerism • Likely monkey-to-monkey variation • Uncertain effects on animal welfare

differences between different approaches for modeling neurological and psychiatric disease using NHPs, including a direct comparison of the pros and cons of transgenic versus interspecies chimeric approaches to disease modeling. A key limitation of the transgenic approach is that our understanding of the genetic roots of certain diseases remains primitive. For example, although some genes responsible for familial Alzheimer's disease are known, the genetic bases for the more common "sporadic" disease are poorly understood.

Indeed, among primates, humans appear to be the only species that manifest the complete clinical and pathological sequela of Alzheimer's disease [31]. Currently, the NHP model that best approximates AD employs aged primates to study changes in brain and behavior associated with AD. However, this model possesses deficiencies. One problem is that while aging primates undergo a cognitive decline, the hippocampus, a memory-associated region, is relatively spared in aged NHPs, despite manifesting severe neuronal loss in AD patients. Having an ApoE4 allele is correlated with earlier onset for AD [32]. One might also pose the possibility of generating a transgenic monkey carrying human ApoE4, a risk allele for AD, and a disease-causing mutation in APP as a potential transgenic model for AD. While such a model will prove useful, it is important to note that the ApoE4 allele has a weak or no obvious effect on AD for individuals from certain ethnicities [33, 34].

To address such insufficiencies, a monkey containing neural tissues derived from AD patients could yield insights into the apparently unique human susceptibility to Alzheimer's disease. Further, by applying a strategy called interspecies blastocyst complementation, one could potentially generate specifically targeted regions of chimeric brains that are entirely human derived [21, 22]. In this method, donor PS cells of species A are injected into the embryos of species B that are organogenesis disabled. If disabling organogenesis is lethal for embryos of species B, the resulting chimera will possess the missing organ completely derived from species A. This strategy has been used to generate rat pancreas in mouse and vice versa and efforts to apply the same strategy to generate human organs in large animals are underway [21, 22]. In theory, one could generate a human stem cell-derived hippocampus by injecting xenogeneic human PS cells into a monkey embryo that is hippocampus disabled. Such an experiment could prove useful for modeling neurological conditions with human-specific biological features. Indeed, the recent development of neural blastocyst complementation in mice suggests the feasibility of translating this approach to human-NHP chimeras [35].

Another example pertains to modeling of Parkinson's disease (PD). The best NHP model of PD uses 1-methyl-4-phenyl-1,2,3,6-tetrahydropyridine (MPTP) to selectively kill dopaminergic neurons, the primary cell type lost in PD [36, 37]. Nonetheless, while the MPTP model reproduces most PD hallmarks, it also has limitations. A major limitation is that the MPTP-based mechanism of cell death differs from the mechanism that occurs in PD—it does not address why people develop PD. Using transgenic NHPs to model diseases with multiple genetic loci containing "risk variants" will be challenging. While there are multiple genes associated with a small percentage of patients with familial variants of PD, most are not. Attempting to model the nonfamilial cases in which multiple genetic loci, especially of small effect, cooperate to predispose to PD will be very difficult using transgenic primates. Therefore, human-monkey chimeras containing tissues derived from PD patients could offer unique advantages compared to both chemical lesion and transgenic based approaches.

Finally, using transgenic NHPs to model psychiatric disorders whose genetic etiology is complex and poorly understood will be challenging [38]. It is noteworthy that NHP models for psychiatric conditions, such as bipolar disorder (BP) and schizophrenia, either do not exist or if existent reveal little about disease pathophysiology. For example, while the phencyclidine (PCP) model for schizophrenia models some of the memory deficits, despite a clear genetic loading, currently available models possess deficiencies [39]. In contrast, using human-monkey chimeras, one could study the function of patient cells carrying multiple genetic risk variants in a humanlike setting. For example, in the case of BP, for which no NHP model exists and for which the genetic bases remain poorly understood, one could generate a human-monkey chimera with neural tissue derived from a patient with bipolar disorder and study the brain and behavior of such an animal. Human-monkey chimeras may prove indispensable for modeling of polygenic diseases.

The creation of human-monkey chimeras raises some additional ethical questions. To those already opposed to monkey research, human-animal chimeras may pose no interesting new issues. But some who accept existing monkey research may worry that a human-monkey chimera would be capable of enhanced suffering, or that it could, by meeting some yet-to-be-specified mental criterion, qualify for special status that would render further experimentation on it unethical. Whether chimerism in portions of the monkey brain would affect cognition or emotion is unknown. The issue is complicated by the fact that no human-monkey chimera can have any chance at life at all, except as a research subject; it may be that, if the chimera has a life that is not too burdensome, there may be fewer objections to its having been created in the lab. Finally, the burden on chimeras of experimentation needs to be weighed against the possible benefits of any given line of research, alternative methods to achieve the same goals, and compared also to the continuing burden of disease, and indeed of experimentation, on humans.

3 A Cautious Path Forward

As new experimental methods for studying aspects of the human brain continue to develop, provocative questions will be raised [40]. Given advances in the stem cell field and chimera research, there exists a need to discuss the scientific merits and ethical concerns that accompany human-monkey chimera formation by scientists, ethicists, and the general public.

In the near term, we recommend that steps be taken to:

1. Support transparent research to study the true nature of chimera-competent and xenogeneic pluripotency in humans and other primates, and to understand the xenogeneic barrier. A step-by-step approach would improve methodology and identify pitfalls. Before making human-monkey chimeras, it will be prudent to first pursue the generation of interspecies between nonhuman primate species. It will be instructive, for example, to introduce xenogeneic PS cells from great apes into monkey host embryos. It will also be useful to investigate the merits of complementary experimental strategies, such as introducing apoptosis-disabled PS cells into preimplantation embryos. Such experiments could inform how to control human PS cell derivative contributions to the chimeric monkey brain.
2. Closely monitor the welfare of all initial human-monkey chimeric models of neurological and psychiatric disease. The transfer of disease-specific human stem cells is likely to affect research animals in ways that compromise rather than enhance their normal capabilities and health [41]; thus all human-monkey neurological chimera research should be independently reviewed and monitored to ensure compliance with appropriate animal welfare standards.
3. Require independent review. A determination must be made that the necessary minimal number of monkeys will be used to answer a meritorious research question for which there exists no reasonable alternative approach.
4. Harmonize differing chimera research guidelines issued by the US National Academy of Sciences (NAS) and the International Society for Stem Cell Research (ISSCR). Although the ISSCR guidelines permit NHP blastocyst complementation experiments pending stem cell ethics review, the NAS guidelines do not.
5. Learn from the example of mitochondrial replacement therapy and human germline editing. Funding agencies should create forums in which experts from the scientific and bioethics communities can deliberate along with the public about risks and

rewards of using such technology. It will be essential to engage the public in addressing the implications of human-monkey chimera formation.

In summary, recent advances in generating chimera-competent PS cells and chimera research raise the prospect of human-monkey chimeras, presenting new possibilities for biomedical and translational research as well as difficult policy challenges. The existence of biologically and clinically relevant differences between human and primates may justify the use of human-primate chimeras to more accurately model human disease. These conditions include but are not limited to Alzheimer's disease and psychiatric disorders such as bipolar disorder and schizophrenia. However, animal welfare considerations call for caution and clear reasoning regarding scientific necessity. Data from these experiments may possibly lead to new ethical arguments against advancing to more complete chimeric experiments. Realizing the promise of human-monkey chimera research in an ethically and scientifically appropriate manner will require a coordinated approach. The field of human-monkey chimera research will need the support of governments, research institutions, and private foundations. If successful, the development of human-monkey chimera technology may expand the breadth of chimera research from the laboratory toward potential clinical benefits for patients with serious neurological and psychiatric disorders.

Acknowledgements

Reprinted from Stem Cells and Development with permission from Mary Ann Liebert, Inc., New Rochelle, NY.

References

1. World Health Organization (2006) Neurological disorders: public health challenges. WHO Press, Switzerland
2. Miller G (2010) Is pharma running out of brainy ideas? Science 329:502–504
3. Takahashi K, Yamanaka S (2006) Induction of pluripotent stem cells from mouse embryonic and adult fibroblast cultures by defined factors. Cell 126:663–676
4. Takahashi K, Tanabe S, Ohnuki M, Narita M, Ichisaka T, Tomoda K, Yamanaka S (2007) Induction of pluripotent stem cells from adult human fibroblasts by defined factors. Cell 131:861–872
5. Dimos JT, Rodolfa KT, Niakan KK, Weisenthal LM, Mitsumoto H, Chung W, Croft GF, Saphier G, Leibel R, Goland R, Wichterle H, Hendersen CE, Eggan K (2008) Induced pluripotent stem cells generated from patients with ALS can be differentiated into motor neurons. Science 321:1218–1221
6. Park IH, Arora N, Huo H, Maherali N, Ahfeldt T, Shimamura A, Lensch MW, Cowan C, Hochedlinger K, Daley GQ (2008) Disease-specific induced pluripotent stem cells. Cell 134:877–886
7. Brennand KJ, Marchetto MC, Benvenisty N, Brustle O, Ebert A, Izpisua Belmonte JC, Kaykas A, Lancaster MA, Livesey FJ, McConnell MJ, McKay RD, Morrow EM, Muotri AR, Panchision DM, Rubin LL, Sawa A, Soldner F, Song H, Studer L, Temple S, Vaccarino FM, Wu J, Vanderhaeghen P, Gage FH, Jaenisch R (2015) Creating patient-specific neural cells for

the in vitro study of brain disorders. Stem Cell Rep 5:933–945

8. Soldner F, Stelzer Y, Shivalila CS, Abraham BJ, Latourelle JC, Barrasa MI, Goldmann J, Myers RH, Young RA, Jaenisch R (2016) Parkinson-associated risk variant in distal enhancer of alpha-synuclein modulates target gene expression. Nature 533:95–99
9. Goldman SA, Nedergaard M, Windrem MS (2015) Modeling cognition and disease using human glial chimeric mice. Glia 63:1483–1493
10. Windrem MS, Osipovitch M, Liu Z, Bates J, Chandler-Militello D, Zou L, Munir J, Schanz S, McCoy K, Miller RH, Wang S, Nedergaard M, Findling RL, Tesar PJ, Goldman S (2017) Human iPSC glial mouse chimeras reveal glial contributions to schizophrenia. Cell Stem Cell 21:195–208
11. Benraiss A, Wang S, Herrlinger S, Li X, Chandler-Militello D, Mauceri J, Burm HB, Toner M, Osipovitch M, Jim Xu Q, Ding F, Wang F, Kang N, Kang J, Curtin PC, Brunner D, Windrem MS, Munoz-Sanjuan I, Nedergaard M, Goldman SA (2016) Human glia can both induce and rescue aspects of disease phenotype in Huntington disease. Nat Commun 7:11758
12. Langston JW, Forno LS, Rebert CS, Irwin I (1984) Selective nigral toxicity after systemic administration of 1-methyl-4phenyl 1-1,2,3,6, tetrahydropyridine (MPTP) in the squirrel monkey. Brain Res 292:390–394
13. Cong L, Ran FA, Cox D, Lin S, Barretto R, Habib N, Hsu PD, Wu X, Jiang W, Marraffini LA, Zhang F (2013) Multiplex genome engineering using CRISPR/Cas systems. Science 339:819–823
14. Mali P, Yang L, Esvelt KM, Aach J, Guell M, DiCarlo JE, Norville JE, Church GM (2013) RNA-guided human genome engineering via Cas9. Science 339:823–826
15. Niu Y, Shen B, Cui Y, Chen Y, Wang J, Wang L, Kang Y, Zhao X, Si W, Li W, Xiang AP, Zhou J, Guo X, Bi Y, Si C, Hu B, Dong G, Wang H, Zhou Z, Li T, Tan T, Pu X, Wang F, Ji S, Zhou Q, Huang X, Ji W, Sha J (2014) Generation of gene-modified cynomolgus monkeys via Cas9/RNA-mediated gene targeting in one-cell embryos. Cell 156:836–845
16. Liu Z, Cai Y, Wang Y, Nie Y, Zhang C, Xu Y, Zhang X, Lu Y, Wang Z, Poo M, Sun Q (2018) Cloning of macaque monkeys by somatic cell nuclear transfer. Cell 172:881–887
17. De Los Angeles A, Ferrari F, Xi R, Fujiwara Y, Benvenisty N, Deng H, Hochedlinger K, Jaenisch R, Lee S, Leitch HG, Lensch MW, Lujan E, Pei D, Rossant J, Wernig M, Park PJ, Daley GQ (2015) Hallmarks of pluripotency. Nature 525:469–478
18. Nichols J, Smith A (2009) Naïve and primed pluripotent states. Cell Stem Cell 4:487–492
19. Tesar PJ, Chenoweth JG, Brook FA, Davies TJ, Evans EP, Mack DL, Gardner RL, McKay RD (2007) New cell lines from mouse epiblast share defining features with human embryonic stem cells. Nature 448:196–199
20. Tachibana M, Sparman M, Ramsey C, Ma H, Lee HS, Penedo MC, Mitalipov S (2012) Generation of chimeric rhesus monkeys. Cell 148:285–295
21. Wu J, Platero-Luengo A, Sakurai M, Sugawara A, Gil MA, Yamauchi T, Suzuki K, Bogliotti YS, Cuello C, Morales Valencia M, Okumura D, Luo J, Vilarino M, Parrilla I, Soto DA, Martinez CA, Hishida T, Sanchez-Bautista S, Martinez-Martinez ML, Wang H, Nohalez A, Aizawa E, Martinez-Redondo P, Ocampo A, Reddy P, Roca J, Maga EA, Esteban CR, Berggren WT, Nunez Delicado E, Lajara J, Guillen I, Guillen P, Campistol JM, Martinez EA, Ross PJ, Izpisua Belmonte JC (2017) Interspecies chimerism with mammalian pluripotent stem cells. Cell 168:473–486
22. Kobayashi T, Yamaguchi T, Hamanaka S, Kato-Itoh M, Yamazaki Y, Ibata M, Sato H, Lee YS, Usui J, Knisely AS, Hirabayashi M, Nakauchi H (2010) Generation of rat pancreas in mouse by interspecific blastocyst injection of pluripotent stem cells. Cell 142:787–799
23. Theunissen TW, Friedli M, He Y, Planet E, O'Neil RC, Markoulaki S, Pontis J, Wang H, Iouranova A, Imbeault M, Duc J, Cohen MA, Wert KJ, Castanon R, Zhang Z, Huang Y, Nery JR, Drotar J, Lungjangwa T, Trono D, Ecker JR, Jaenisch R (2016) Molecular criteria for defining the naïve human pluripotent state. Cell Stem Cell 19:502–515
24. Cohen MA, Markoulaki S, Jaenisch R (2018) Matched developmental timing of donor cells with the host is crucial for chimera formation. Stem Cell Rep 10:1–8
25. Wu J et al (2015) An alternative pluripotent state confers interspecies chimeric competency. Nature 521:316–321
26. Mascetti VL, Pedersen RA (2016) Human-mouse chimerism validates human stem cell pluripotency. Cell Stem Cell 18:67–72
27. Boroviak T, Nichols J (2017) Primate embryogenesis predicts the hallmarks of human naïve pluripotency. Development 144:175–186
28. Ramos-Ibeas P, Sang F, Zhu Q, Tang WWC, Withey S, Klisch D, Loose M, Surani MA, Alberio R. Lineage segregation, pluripotency and X-chromosome inactivation in the pig

pre-gastrulation embryo. bioRxiv. Accessed 30 Jun 2018

29. Tsukiyama T, Ohinata Y (2014) A modified EpiSC culture condition containing a GSK3 inhibitor can support germline-competent pluripotency in mice. PLoS One 9:e95329

30. De Los Angeles A, Pho N, Redmond DE Jr (2018) Generating human organs via interspecies chimera formation: advances and barriers. Yale J Biol Med 91:333–342

31. Walker LC, Jucker M (2017) The exceptional vulnerability of humans to Alzheimer's disease. Trends Mol Med 23:534–545

32. Gomez-Isla T, West HL, Rebeck GW, Harr SD, Growdon JH, Locascio JJ, Perls TT, Lipsitz LA, Hyman BT (1996) Clinical and pathological correlates of apolipoprotein E epsilon 4 in Alzheimer's disease. Ann Neurol 39:62–70

33. Farrer LA, Cupples LA, Haines JL, Hyman B, Kukull WA, Mayeux R, Myers RH, Pericak-Vance MA, Risch N, van Duijin CM (1997) Effects of age, sex, and ethnicity on the association between apolipoprotein E genotype and Alzheimer disease. A meta-analysis. APOE and Alzheimer disease meta-analysis consortium. JAMA 278:1349–1356

34. Tang MX, Stern Y, Marder K, Bell K, Gurland B, Lantigua R, Andrews H, Feng L, Tycko B, Mayeux R (1998) The APOE-epsilon4 allele and the risk of Alzheimer disease among African Americans, whites, and Hispanics. JAMA 279:751–755

35. Chang AN, Liang Z, Dai H-Q, Chapdelaine-Williams AM, Andrews N, Bronson RT, Schwer B, Alt FW (2018) Neural blastocyst complementation enables mouse forebrain organogenesis. Nature 563:126–130

36. Burns RS, Chiueh CC, Markey SP, Ebert MH, Jacobwitz DM, Kopin IJ (1983) A primate model of Parkinsonism: selective destruction of dopaminergic neurons in the pars compacta of the substantia nigra by N-methyl-4-phenyl-1,2,3,6,-tetrahydropyridine. PNAS 80:4546–4550

37. Langston JW, Ballard P, Tetrud JW, Irwin I (1983) Chronic parkinsonism in humans due to a product of meperidine-analog synthesis. Science 219:979–980

38. Hyman SE (2018) The daunting polygenicity of mental illness: making a new map. Philos Trans R Soc Lond Ser B Biol Sci 19:373

39. Jentsch JD, Redmond DE Jr, Elsworth JD, Taylor JR, Youngren KD, Roth RH (1997) Enduring cognitive deficits and cortical dopamine dysfunction in monkeys after long-term administration of phencyclidine. Science 277:953–955

40. Farahany NA, Greely HT, Hyman S, Koch C, Grady C, Pasca SP, Sestan N, Arlotta P, Bernat JL, Ting J, Lunshof JE, Iyer EPR, Hyun I, Capestany BH, Church GM, Huang H, Song H (2018) The ethics of experimenting with human brain tissue. Nature 556:429–432

41. Hyun I (2016) Illusory fears must not stifle chimaera research. Nature 537:281

Index

Insoo Hyun and Alejandro De Los Angeles (eds.), *Chimera Research: Methods and Protocols*, Methods in Molecular Biology, vol. 2005, https://doi.org/10.1007/978-1-4939-9524-0,

Zeitfracht Medien GmbH
Ferdinand-Jühlke-Straße 7
99095 Erfurt, Deutschland
produktsicherheit@kolibri360.de